A guide
birds
of Anguilla

by Steve H Holliday, Karim V D Hodge and Damien E Hughes

Published by
The Royal Society for the Protection of Birds
The Lodge, Sandy, Bedfordshire SG19 2DL, England

10-digit ISBN: 1 905601 10 7   13-digit ISBN: 978 905601 10 3

Designed by
NatureBureau, Newbury, UK

Printed by
Information Press, Oxford, UK

All images are by Gillian E Holliday except for the following:

Steve H Holliday, pages 12, 14, 17 (lower), 19 (lower), 20, 21, 25, 28, 47, 75, 80, 82, 100
David Norton (rspb-images.com), page 31
Chris Gomersall (rspb-images.com), page 32
S & D & K Maslowski/FLPA, page 33
Tom and Pam Gardner/FLPA, page 34
Richard Revels (rspb-images.com), page 35
Flip de Nooyer/Foto Natura/FLPA, page 37
Malcolm Schuyl/FLPA, pages 49, 58
David Kjaer (naturepl.com), page 67
Tom Vezo/Minden Pictures/FLPA, pages 68, 83
Tony Hamblin (rspb-images.com), page 78

Contents

Foreword

From the Chief Minister of Anguilla

Congratulations to the Anguilla National Trust for an excellent job in publishing yet another informative and attractive book on Anguilla's natural environment. This is the fourth publication by the Trust and together they serve to foster an appreciation of Anguilla's fragile natural resources and enhance awareness of conservation issues that continue to affect the development of the island.

The Government of Anguilla fully supports this and all other publications celebrating Anguilla's natural environment. These guides now provide new areas of interest for our tourism market and are proving critical to our island's tourism diversification strategy. It is against this backdrop that the Government will continue to fully address and implement its commitments under the United Kingdom Environment Charter Agreement as well as its obligations under the St. Georges Declaration of Principles of Environmental Sustainability.

The Anguilla National Environmental Management Strategy (NEMS) is the key policy document to implement our environmental programmes. The Government of Anguilla recognises the continued and co-operative relationship with the National Trust and values the significant contribution the Trust has made towards national sustainable development and the long-term management of the island's environment.

The Government also recognises that Anguilla's natural resources extend beyond the beaches and inshore waters to other habitats such as mangroves, ponds, forest, scrub and the cays and that our commitment through the NEMS comes at a time of growing interest in wildlife tourism which could extend the tourism season. Finally, the Government of Anguilla encourages residents and visitors to use this excellent guide, and to support the Anguilla National Trust and its work.

I sincerely hope this book gives you a glimpse into the heart of Anguilla, helps you enjoy our biodiversity wrapped in blue and enables you to truly live the Anguillian experience of feeling is believing.

Hon. Mr. Osbourne B. Fleming
Chief Minister and Minister of Environment

From Andrew George, Governor of Anguilla

I am very pleased to have been asked to contribute to this important publication on the birds of Anguilla. In common with other UK Overseas Territories, Anguilla is a rich habitat for many species of animals, and makes an irreplaceable contribution to the planet's biodiversity. The UK Government, through the Overseas Territories Environment Programme, promotes the conservation and protection of these vital resources as a fundamental element in the process of sustainable development.

The production of this guide is an outstanding example of collaborative effort between the Government of Anguilla, the Anguilla National Trust, the UK Government, and the Royal Society for the Protection of Birds. The guide will inform and inspire both the people of Anguilla themselves and many actual and potential visitors to cherish the island's rich natural heritage. All proceeds from sales of the guide will go to support the work of the Anguilla National Trust.

Andrew George, Governor of Anguilla

Acknowledgements

We would like to thank His Excellency Andrew George, Governor of Anguilla, and the Honourable Mr Osbourne B Fleming, Anguilla's Chief Minister for kindly providing a foreword to this guide.

Two people in particular have made a significant contribution to our knowledge and understanding of birds on Anguilla. Sir Emile Gumbs who has perhaps the longest experience of birdwatching on Anguilla, provided an invaluable historical perspective to this guide, and shared many of his bird sightings particularly from his local site of Road Salt Pond. Roy Thomas collected his own extensive data on the birds of Anguilla together with records by resident and visiting birdwatchers and published the first book on the birds of the islands in 1996. Roy's book provides the majority of bird records up to 1996.

This guide has been made possible by the joint efforts of the Anguilla National Trust (ANT) and staff from the Royal Society for the Protection of Birds (RSPB) in the UK, working in partnership since 1999. From the RSPB, this partnership work was initially developed and guided by Jim Stevenson and continued by Sarah Sanders. David Pritchard provided a report on the nature conservation context and Tony Murray, Melanie Bryer, Ian Fisher and Julian Hughes made significant contributions to background research and bird data. Helen Franklin and Rachel Roberts provided help with data management. From the ANT, Ijahnya Christian was instrumental in setting up the initial project and we have valued the support of Gina Brooks, Avon Carty, Oliver Hodge and all involved with the Trust. We have particularly appreciated Sarah Sanders' contribution to the editorial team and for contributing the chapter on Important Bird Areas of Anguilla. Farah Mukida, Natalia Collier, Adam Brown, Julian Hughes and Gillian Holliday have made helpful comments on the text. Charlotte and Lucy Holliday helped research the site information and directions, Colin Wilkinson produced the maps, Mike Pollard scanned in images and Ian Lewington helped to confirm recent rare bird identification. We are indebted to Gillian Holliday for providing the majority of bird images for the identification section and for taking all these on the island. At the editing stage Mandy Carter and Lisa Fell from the RSPB made valuable contributions as did Peter Creed from our designers Naturebureau.

The financial support of the UK Government through its Overseas Territories Environment Programme has been vital to the project and the support of Department for International Development and Foreign and Commonwealth Office staff Dick Beales, Peter Johnstone, Alan Huckle, Andrew George, Jackie Barlow and Joe Legg has been greatly appreciated.

Ultimately guides such as this are based on the enthusiasm, fieldwork and bird notes from many observers and we would particularly like to thank the following (with apologies for any omissions): Amanda Adams, Candis Adams, Willie Allan, Ken Banks, Adam Brown, Hans von Buel, Avon Carty, Brenda Carty, Natalia Collier, Jacqueline Connor, Rhon Connor, Steve Donahue, Suzan Donahue, Kimberley Fleming, Ingrid Fullington, Sir Emile Gumbs, James Gumbs, Rasheem Gumbs, Calyn Harrigan, Charleon Hodge, Jarmella Hodge, Marslyn Hodge, Oliver Hodge, Michael Kaprowski, Penny Legg, John Lloyd, Frances Marks, Richard Moore, Farah Mukhida, Bob O'Keefe, Mary O'Keefe, Frances Marks, Colville Petty, Ann Pienkowski, Mike Pienkowski, Jonathan Randall, Mitchell Randall, Lanny Richardson, Jarmarie Richardson and Bonny Warner.

About this guide

Anguilla is a wonderful place to watch birds and wildlife. Tropical birds from the Caribbean and migratory birds from North America can be found together on these small islands. This book aims to be a helpful introduction and reference guide to these birds and provides information to help you find and identify all the species of birds commonly found on Anguilla. It also includes information on all the birds seen on Anguilla and the smaller islands in the archipelago between 1990 and 2006.

The guide should be of interest to both the casual and keen birdwatcher and to local students and teachers. If you live on the island you will see the majority of the species illustrated as you travel around the island. For visitors, there are many attractions on this beautiful island and we hope you will take time out to visit some of the sites we have included and to enjoy and learn about Anguilla's spectacular birdlife.

The birdwatching notes are written from a purely local perspective although much of the information will be of help on visits to some neighbouring islands in the Lesser Antilles. The guide includes information on how to separate some species that are difficult to identify but the text is not comprehensive. If you are keen to identify some of the more tricky species, or to age individual birds, we recommend you use this guide together with a field guide. To date there has been limited study on the breeding behaviour of birds on Anguilla, other than the extent of breeding seasons, and more information on this aspect is available in some regional bird guides.

This guide is a joint collaboration of the Anguilla National Trust (ANT) and the RSPB. All proceeds from the sale of this guide will go to support the work of the ANT. Please help the Trust further in its work by becoming a member, taking part in its activities and participating in its survey and monitoring programmes. You could help to provide information to help the conservation and management of important sites and habitats and add to our knowledge of Anguilla's birds and other wildlife for future editions of this guide.

If this guide encourages you to get involved, to visit some of the island's less known areas and to share in the authors' passion for the birds of Anguilla we will consider it a great success.

Road Bay, Sandy Ground village and Road Salt Pond

Introduction

From the time Amerindians first arrived on Anguilla in the distant past, the island's wildlife and rich natural resources have played a vital part in island life. The forest and scrub covering the low rocky islands were a source of food and provided shelter from the hot sunny conditions and constant breezes drawn in from the Atlantic Ocean. The beautiful beaches were ideal to launch boats out to fish in the bays and coral reefs where feeding flocks of brown boobies and royal terns provided a guide to where fish were shoaling. In the skies above sheltered bays such as Road Bay, they would have looked up to see magnificent frigatebirds hanging like dark angular kites watching out for seabirds with prey to pursue. Springs and pools provided essential sources of fresh water and occasional storm surges would have flooded coastal lagoons whose shallow waters would have teemed with water birds such as egrets and a host of migratory ducks and shore birds flying in from North America.

Anguilla has seen many changes in the intervening centuries but in many ways the island's natural resources still determine the pattern of life and development. Fishing remains important and local fishermen provide the restaurants with some of the freshest seafood possible. The stunningly beautiful beaches and inshore coral reefs attract locals and tourists to swim, snorkel and dive in the clear seas. The spectacular view from the hill across the salt pond into Road Bay now has the settlement of Sandy Ground behind the beach with its vibrant beach atmosphere, yet the frigatebirds are still there and for many people are an essential part of the scene and experience in the tropical Caribbean.

Each year, as they have done for thousands of years, water birds from as far away as Arctic Canada continue to pour out of North America down the West Atlantic Flyway, to spend the northern winter on Anguilla's wetlands or to use them as feeding stops on even longer journeys to South America. We are now mirroring these journeys as air travel enables increasing numbers of visitors to head south making the island a popular holiday destination.

The attraction of Anguilla is captured in some early publicity for the island, *Tranquillity wrapped in blue*. It conjures up pictures of beautiful stretches of sandy beach curving around clear, blue tropical seas. These wonderful inshore waters hold coral reefs rich in fish and other marine life. The mainland and the cays have low but ruggedly attractive limestone and sandstone cliffs. Ashore, the island is clothed in low scrub and woodland, a rich and varied habitat much underestimated for its value. Some shallow soils have formed, providing the basis for a small amount of agriculture. There is also a range of natural ponds – from coastal saline lagoons, many adapted for a former salt industry, to seasonal ponds and a few fed by freshwater springs.

These natural habitats continue to support a wide range of plants and animals. They help make the islands a wonderful place for wildlife and in particular a wonderful place for birds. This guide is the fouth in a series exploring Anguilla's wildlife and aims to encourage a wider interest and enjoyment of birdlife by residents, teachers, students and visitors. It follows recent publications on wetlands, reptiles and amphibians and one on common plants.

Brown pelican and royal terns

Anguilla's position in the extreme north-eastern corner of the Lesser Antilles, where the Caribbean Sea meets the Atlantic Ocean, influences both its character and its current birdlife. Marine and inshore fish populations will always have provided a rich food source for seabirds. The mainland and the smaller islands and cays have provided safe nesting sites for seabirds for many thousands of years. Seabirds are now all but lost on the mainland, but recent studies have shown that islands like Dog Island and Sombrero hold some of the most important seabird colonies in the whole Caribbean region.

The softer coastal features of beaches, sand dunes and coastal lagoons provide nesting and feeding areas for resident and migratory water birds such as herons, egrets, ducks and shore birds. The shallow ponds provide ideal conditions for a range of midges and flies to breed and can teem with larvae and flying insects in incredibly productive natural systems. The birds in turn arrive in large numbers to feed on these invertebrates.

The small size and restricted height of the mainland and its cays limits the range of vegetation to lowland dry forest and scrub. Away from sheltered areas extensive areas of low scrub grows, adapted to the effects of strong, drying winds and salt spray. The vegetation has limited the range and populations of land birds on Anguilla although it still holds several species restricted to the Lesser Antilles such as the green-throated carib and the Lesser Antillean bullfinch. The woodland also attracts a small number of migratory land birds from North America.

Bird populations are found throughout the island and its cays. Residents and visitors alike will be aware of the role birds play as part of the whole island experience. Pelicans, brown boobies, and terns fish alongside fishing and sailing boats. The seasonal arrivals of kingbirds (the chincherry) are well known as they hunt for insects over the open ground around dwellings. The soft calls of the familiar Zenaida dove are a background sound to island life and its familiarity led to Anguillians voting it their national bird in 1993.

Amid the stabilities of life on the island such as the settled population, village communities, traditions such as boat racing and the constant sunshine and breezes, the Caribbean area also has a dynamic nature. Periodic hurricanes have a dramatic impact on people, property, wildlife and natural resources. They can change sandy coastlines, kill or damage trees and mangroves and displace wildlife. It can take many years for people and the land to recover.

Parts of the island's natural resources are still recovering from the impacts of Hurricanes Luis and Marilyn in 1995 and Hurricane Lenny in 1999. Emblematic of effects on wildlife was the disappearance of both native hummingbirds after the hurricane in 1995. By the year 2000 a few green-throated caribs had returned from nearby islands and were again breeding. The smaller Antillean crested hummingbird has taken longer to return with the first sighting as late as 2001 and it was 2004 before more than a single bird was seen. There is still no evidence of breeding a decade after their initial loss.

Just as the path of hurricanes can be unpredictable, the direction and frequency of other weather systems varies. The period August–December can see considerable rainfall or an extended dry period. In addition to the pressure this places on water supplies, water levels in the ponds can also fluctuate. The shallower seasonal ponds such as Blowing Point Pond or East End Pond can dry out completely. If this happens numbers of wetland birds can be much reduced as shown by the regular wetland bird counts carried out by the Anguilla National Trust (ANT). There is evidence that wetland birds will move around islands in the Lesser Antilles in response to food availability and water levels in the ponds although information on these movements is limited. In much the same way, Anguillians have moved around the islands at times in search of trade or employment.

As people have moved around the Caribbean they have brought with them a range of animals as pets or livestock. These movements and associated trade have also brought unwanted arrivals such as rats. The more settled islands in the Caribbean now have populations of predators and feral animals. On Anguilla, populations of stray goats, cats and dogs, together with introduced rats, have all had a detrimental impact on native wildlife. The only breeding seabirds left on the mainland are a few cliff-nesting tropicbirds and scattered colonies of least terns. The internationally important seabird colonies on Anguilla are now all on uninhabited offshore islands, although some of these are under threat from goats and particularly rats, which are now known to be present on at least Dog Island and Prickly Pear East.

Perhaps the biggest change currently underway is the continuing increase in tourism and visitors to the island. This has been largely concentrated on Anguilla's coastal resources with new resorts and hotels built to take advantage of the beaches, clear seas and the tranquillity. The challenge in a growing market is to develop without compromising the environment that local people enjoy and visitors appreciate. The new millennium has seen a considerable increase in pressure on land for development and the fragile habitats, important to wildlife, will need wise stewardship from government if birds and other wildlife are to thrive.

Much of the pressure is coming from increased economic growth and corresponds to the large amount of foreign direct investment in the tourism sector. In just a few short decades, development pressures have grown on coastal and wetland sites threatening vital links in a migration chain that has sustained birds for millennia and inspired people for generations. Ironically, the best times for seeing birds on Anguilla are at times that could extend the traditional tourism season. From April to early June and August through to November resident birds can be seen alongside migrating birds at the peak of the migration season. Although some loss of habitat is inevitable in the current development phase, it is possible to develop further in a way that ensures Anguilla's rich natural resources continue to support local people, visitors and wildlife. Birds are great indicators of the health of the environment and the bird monitoring carried out by ANT will enable the island's development to be tested for its sustainability.

Cattle egret

Government is responding to these development challenges. It is implementing a joint Environment Charter with the UK, as well as the St Georges Declaration (SGD) for environmental sustainability in the Organisation of Eastern Caribbean States, and has adopted a National Environment Management Strategy. Significant strides are being made towards strengthening the environmental legislative framework to enable it to set its national workplan within a regional and international context. Government is also developing the skills and capability of staff members in a range of departments directly involved in environmental conservation work.

Together these initiatives will set a national framework for environmental management and sustainable development and play a vital role in ensuring Anguilla's wonderful natural environment is safeguarded and managed for future generations. Continued support of the ANT by government and an active membership will also enable essential conservation work to be carried out.

Imagine the island without the noisy colonies of stilts enlivening the edges of the ponds, of a coastal skyline without the man-of-war or frigatebird, hanging in the tropical skies. How much poorer will be the catalogue of holiday memories of visitors without the experience of watching pelicans plunge into sea or enjoying a glimpse of the iridescent green of a hummingbird? These have been very much part of the island scene since the Amerindians inhabited the island they called Malliouhana. Let us hope they continue to remain part of the future of what we know as Anguilla.

Rendezvous Bay

Anguilla: an introduction

Location

The Anguilla archipelago of islands is bordered by two great bodies of water, the Atlantic Ocean to the north and east and the Caribbean Sea to the south and west. Located at 18°30'N, 63°50'W, it is the most northerly part of the Leeward Islands set within the Lesser Antilles. The archipelago is composed of Anguilla and several offshore islets or cays. The larger ones, that are more than merely emergent rocks, include mainland Anguilla, Anguillita, Scilly Cay, Scrub Island, Little Scrub Island, Dog Island, West Cay (off Dog Island), Mid Cay (off Dog Island), East Cay (off Dog Island), Sandy Island, Seal Island, Prickly Pear Cay West, Prickly Pear Cay East and Sombrero, which sits on its own island bank.

The people and culture

The friendly people of Anguilla help contribute to its reputation as one of the safest, most relaxed tourism destinations in the Caribbean, where solitude and tranquility are promoted as special qualities. However, there is a growing realisation that a development strategy based solely on economic development, including growing levels of tourism, endangers the wealth of knowledge, values and traditions Anguillians collectively call their heritage.

The estimated population size of Anguilla in 2001 was 11,561 (according to the census). This population level leaves the island relatively sparsely settled when compared with some neighbouring islands and most villages on Anguilla have large tracts of undeveloped land around them. The population distribution is at its highest concentration among the villages of The Valley, South Hill, North Side, The Quarter and Stoney Ground.

Many of Anguilla's cultural traditions are embedded in the art of boat racing, which is the national sport. Given the island's long British links it must be one of the few places where cricket takes a back seat to an indigenous sport! Boat racing is a unique sport that is linked to the island's economic history where, after the collapse of commercial agriculture, there was a need for transport to facilitate large scale movements of workers and to trade with other islands. The boat racing today is a direct descendant from these activities and the large sailed boats are based on original fishing boats. Boat racing is now part of the activities of every national holiday and is most memorable during the week long festivities in August that also include a local carnival.

Current economic situation

After many years with an essentially subsistence economy, Anguilla has developed rapidly over the last 20 years or so, due almost entirely to the expansion of the tourism industry on the island. From a base of 8,172 visitor arrivals in 1980, numbers have averaged at over 100,000 visitors a year since 1991. The Gross Domestic Product (GDP) has more than tripled to EC$240 million (US$89) over the last 15 years with a rise in average household and per capita incomes. As a direct result of this increase in economic activity, the population has increased by over 70% since 1982 from 6,680 to 11,561 (in 2001) compared to an increase of barely 1,000 in the preceding 20 years. A large part of this increase has been due to in-migration, mainly from other Caribbean countries and these new arrivals now make up 28% of the population.

The economy has diversified in recent years particularly into banking, insurance and communications leaving the primary activities of agriculture and fishing providing a 3% share of economic activity. Tourism, with all its direct activity and indirect activity in support services, remains by far the biggest source of economic generation. This increase highlights the island's dependency on the tourism sector which currently provides 60% of GDP and 48% of employment. This dependence makes the island highly vulnerable to trends in international and regional tourism.

Structure, climate and natural resources

Anguilla's surface features are unique within the Lesser Antillean islands. Most of the other islands are mountainous resulting in higher rainfall. In contrast, Anguilla is low-lying and dry with a maximum elevation of 65 m. A few sandstone outcrops occur such as at Crocus Bay, West End and on Dog Island but most of the island's structure is of a porous karstic limestone structure or weathered coral that allows water to seep in rather than accumulate on the surface. As a result, surface waters are invariably saline. The island's low-lying coastline leaves it especially vulnerable to potential impacts of climate change such as sea-level rise.

Anguilla lies within the tropics with a wet season extending from June to November that coincides with the Atlantic hurricane season. The majority of an entire year's 900–1,000 mm of rain

Boat racing – the national sport

may fall in only a few weeks in this period and may cause temporary, localised flooding in low-lying areas. The dry season on the other hand extends from December to May. During this period, occasional light showers may fall, but intense rainfall is rare. Several pockets of rich, red soils occur providing the basis for agriculture and these can be seen in The Valley, Low Ground near Limestone Bay and The Welches for example.

The annual average rainfall for the island is about 970 mm with an annual range from 460 mm to over 2,050 mm depending on the number and intensity of storms that pass through Anguilla. The minimum monthly rainfall can be less than 5 mm for any given month although these are usually within the dry period. The maximum one-day rainfall can be as much as 360 mm (Data from Department of Agriculture Station in The Valley). The average monthly, ambient temperature ranges from 26.4°C to 29.8°C (79.5–85.6°F). Easterly winds are predominant, blowing at about 15 to 28 km/h for most of the year. Winds can also blow from the north-east, east, south-east and occasionally from the south. The island is periodically hit by hurricanes and tropical storms. Even hurricanes and storms that pass near to the island can cause some wind damage and result in torrential rain.

Wildlife habitats and landscape

The islands contain a rich variety of natural habitats from dazzling sea grass beds and coral reefs to dry, evergreen woodland and forest including mangroves and grassland areas. The broad habitats featured in this guide are the dry forest and scrub, brackish and freshwater ponds, and the marine environment from coastal cliffs and beaches to the seas beyond.

The dry woodland forest and the grasslands of Anguilla provide a mosaic of habitats, a rich tapestry within an island ecosystem. Although limited by the islands' small size and low elevation it is very diverse and the range of micro-habitats within it are home to a variety of animals and over 300 species of native plants. Even with the limited elevations to be found on the island there are a number of vantage points to see the sweeps of wind-sculptured scrub and woodland especially on the south-eastern coast. Among the native reptiles are the Anguillan bank racer, a small non-poisonous snake that hunts for prey within the forest, and the Sombrero black ground lizard which is found only on this remote rock outcrop and nowhere else in the world. There is also a large native land iguana, now found on only a few islands in the eastern Caribbean, which is critically endangered and on the verge of local extinction. It faces threats from human development pressures to its habitat, predation from stray cats and dogs, and competition for food from feral goats. This charismatic species needs urgent action to save it within its limited current range and to ensure that its ecological role is not lost to the island.

The native woody habitat has evolved to cope with a harsh environment of continuous winds, salt spray, unpredictable rainfall and thin soils that retain little water. This has led to many tough, weathered trees and bushes with hard evergreen leaves. The vegetation also has to withstand the pressures and trauma that tropical storms and occasional hurricanes inflict on the islands. Nature's clock occasionally has to be reset after a hurricane event and it can take many months or even years for the vegetation and wildlife to recover in some ecologically sensitive habitats.

Mangroves

The salt tolerant mangroves are a vital and naturally occurring part of Anguilla's wetland systems. These are currently in a recovery phase following severe hurricane damage in the 1990s. Pressure on these systems has been further increased by a loss of many areas over the years to tourism and infrastructure development. None-the-less the remaining mangroves occur in a range of ecologically diverse wetlands where buttonwood predominates and where on some ponds, four species of mangrove are found. Mangroves provide the island and its world class beaches with protection from wave action and erosion as well as providing a nursery for fish, conches and lobsters among their roots. Their dense network of branches are home to pond crabs and provide foraging or nest sites for a range of land and water birds such as the yellow-crowned night-heron, snowy egret, white-cheeked pintail and black-necked stilt.

These ponds and wetlands are incredibly productive natural systems and provide insect food and a habitat for large numbers of breeding water birds and a host of migratory birds that stream down from North America to spend the northern winter in the Caribbean or South America. Many of the shore birds breed in the remote High Arctic and within weeks of finishing their short breeding season they can be seen at close range feeding on Anguilla's wetlands. The ponds also attract a range of migratory ducks, ospreys, belted kingfishers and several species of migratory warblers that find insect prey in the surrounding vegetation.

The beaches and inshore waters of the islands are justifiably world-renowned yet they remain fragile and vulnerable natural systems. The beaches provide excellent areas for public enjoyment yet our enjoyment brings competition for three species of sea turtles that need undisturbed habitat to lay their eggs. The three species have different niches with the hawksbill turtle nesting up in the sea grapes, the green turtle in the vegetation line on the beach and the huge leatherback turtle nesting on the beach itself. In the clear shallow inshore waters, the sea grass beds provide a very rich browsing and grazing environment for the green turtles which feed solely on sea or turtle grass that is most evident in Little Bay. The reef systems around Scrub Island, the Prickly Pear cays and Dog Island provide the hawksbill turtles with rich sources of sponges, coral tissues and other food. A range of pale fish can be found over sandy inshore waters along with occasional starfish and a range of beautiful shells washed in before they are stranded on the beach tide line.

The rocky shores and coral reefs are more diverse and although they show signs of past damage they are home to a dazzling array of fish, corals and other creatures. It is easy to see 20 or more species of fish while snorkelling close to the beach at Little Bay or Shoal Bay East for example, with groupers, parrot fish, pipe fish and wrasse among the common groups.

Overview of biodiversity

For the size of its land area there is no doubt that Anguilla is rich in biological diversity. Over 300 native species of plant, 135 species of bird, 21 species of reptile and five species of bat have been recorded here. Many resident species have a restricted range within the Caribbean and a few are endemic species, found only on Anguilla and nowhere else in the world. These include one endemic plant, the spiny *Rondaletia* or Anguilla bush, and two lizards, the Sombrero black lizard and the Little Scrub ground lizard. A recent visit to Sombrero found over 40 endemic species of insect further demonstrating the unique nature of this tiny rock outcrop.

Little Bay

Many groups of plants and animals such as insects, fungi and lichens have been little studied on the islands and there is much left to discover. Recent publications on plants, reptiles, amphibians and now birds continue to extend our knowledge. A number of bird surveys carried out by the ANT and the RSPB and others have shown Anguilla to be internationally important for its seabirds and these are highlighted in a later chapter on Important Bird Areas.

The movement of people around the Caribbean and current levels of tourism and international trade have brought in a range of introduced species, some intentional others accidental. Over 200 species of plant have been introduced by man. The Amerindians brought in crops such as corn and cassava while later settlers and plantation owners brought in cotton, citrus fruits, tamarinds and pomme-surettes. The interest in imported plants for landscaping and gardens, mostly associated with tourism development, has seen a number of unwanted arrivals in the plant pots that are already causing problems for native plants and animals. These include a land snail and several species of ants, including fire ants.

Conserving biodiversity – current pressures and future challenges

On Anguilla, like on most of the neighbouring small-island developing states in the Caribbean and globally, it is clear that biodiversity, the range of plants and animals, is under more serious threat than ever before. A surge in development for housing and tourism related activities and an increase in population growth on the island and from immigration have placed severe pressures on an increasingly stressed environment. These pressures will be exacerbated by any impacts of climate change.

The main problems facing biodiversity conservation on Anguilla are primarily habitat loss due to an increase in economic activity and a lack of public awareness and appreciation about the importance of the environment to the island. Despite the many ecological services Anguilla's dry evergreen forest, wetlands and coral reefs provide, many are being destroyed or compromised for the sake of economic development as no comprehensive environmental legislation is in place that would ensure their conservation. With 95% of land under private ownership, Government faces a difficult challenge to establish a Protected Area System.

Other threats that are less widespread but capable of creating similar environmental problems include introduced (alien) species, such as the green iguana which may cause extinction of the native iguana, and disturbance to important breeding or feeding areas for a range of species. There are also potential threats to the high quality environment of the islands from potential pollution incidents and the growth in domestic and industrial waste from high levels of tourism and its related activities.

Although some groups of plants and animals have been surveyed there are still huge gaps in our knowledge. The ecology of the Anguillian archipelago needs to be assessed to enable its complex functioning to be better understood. Recent visits to Sombrero suggest there are species remaining, particularly insects, that have yet to be named by science. Biodiversity is important to us in many ways and the value of these species to human existence as potential sources for medicines or as vital ecosystem regulators have barely, if at all, been studied. Anguilla is at the early stages of its biodiversity planning: identifying what species and habitats are present and where; setting targets to maintain or restore biodiversity; which species, sites and habitats are important regionally and internationally; which are most threatened and where is action most urgently needed.

There is no doubt that Anguilla has already lost significant parts of its diversity of life through the lack of a concerted and carefully planned environmental strategy where natural resources are considered as part of a sustainable approach to growth. The resources to support actions appear to be scarce at present, but with the right political leadership and public support much can be achieved. One message keeps returning as the island manages the current high levels of development. Tourism is a major contributor to Anguilla's economy. However, if not carefully managed, it is in danger of destroying the natural assets on which it is based and ultimately depends upon.

Next steps

By buying this guide, and other guides to Anguilla's biodiversity, you are supporting the on-going efforts of the ANT to conserve the nation's rich natural heritage. This is both through a direct contribution of funds and hopefully, by learning more about the importance of the island's birds and the wetlands, forest, scrub and marine environments they depend upon and the need for their conservation.

American golden plover

Ultimately the conservation of these fantastic natural resources and the spectacular birdlife lies in the hands of the people of this small island nation. A strong political will is needed to achieve a proper level of legislation to ably protect even the most vulnerable species and their habitats. Serious political action is dependent above all on an environmentally aware public and there is a great need for an ongoing conservation education and awareness programmes and activities, both formal through the education system and informally through the ANT's heritage tours and other public events. Given the importance of a high quality environment to Anguilla's economic and social development, its role in sustainable development must become a recurrent theme in the education curricula of the nation's pre, primary and comprehensive education systems. The ANT is actively playing a role in raising awareness of environmental issues and is increasingly seeking public input and involvement in a growing effort to ensure the survival of Anguilla's natural heritage.

Finally, the acquisition of knowledge about the biology of native species is equally important as education and legislation. Only a few species have been studied extensively, and long term studies exist only for ground lizards. The ANT, in co-operation with international conservation and funding agencies, continues to work with Government to develop conservation plans for native species, including Anguilla's birds, native iguanas, the racer snake, lizards of Little Scrub and Sombrero and the sea turtles.

Addressing conservation problems is never easy and is even more challenging on an island nation where land is limited and mostly in private ownership. However, in Anguilla the main key is that of education and an awareness that a healthy natural environment rich in birds and biodiversity is in everyone's long term interest. This will be an essential element if any activity or project is to be truly successful and future generations and future visitors are able to enjoy the dazzling array of plants, animals and other natural resources that we find today.

East End Pond – viewing platform

The Anguilla National Trust

The Anguilla National Trust was established by law in 1988 as the nation's custodian on matters of natural, cultural, archaeological, and environmental importance. As a statutory body, the Trust was commissioned to 'promote the permanent preservation for the benefit of Anguillians of lands of beauty… the preservation (so far as possible) of their natural aspect features and animal and plant life'. Additionally, it was expected to 'maintain and manage lands whether or not acquired by the Trust as open spaces or places of public resort… for the purposes of public recreation, resort or instruction'.

Almost two decades later, the protection of Anguilla's national heritage has remained the focal point of the Trust's work. Over this time it has worked with a growing number of stakeholders and partners on a range of collaborative projects that help further its aims. The Trust has maintained particularly close working relationships with the Government of Anguilla in their efforts to implement national, regional and international environmental policy. These policies include the UK White Paper on Partnership for Progress and Prosperity, St Georges Declaration, the National Environmental Strategy, the United Kingdom Overseas Territories Environment Charter and the ruling Anguilla United Front Party's Manifesto.

The programmes of work and activities carried out by the Trust have been made possible through considerable financial support in the form of an annual grant from the Government of Anguilla. This has been augmented by external grant support from agencies such as the United Nations Development Programme, Canadian International Development Agency, Organisation of Eastern Caribbean States-Environment & Sustainable Development Unit, the Royal Society for the Protection of Birds (RSPB), and the new joint DFID/FCO Environment Programme OTEP, as well as the Caribbean Natural Resources Institute. Historically, these programmes have focused on particular plants and animals.

The Trust has carried out work to protect the four species of marine turtles found in Anguilla's waters and, more specifically those nesting on Anguilla's beaches. The organisation's work on reptiles has focused on protecting and increasing the numbers of the endangered Lesser Antillean iguana and Anguillian racer snake. This work is not confined to the mainland but has included efforts towards the conservation of the endemic ground lizards found on Sombrero and Little Scrub. This work recently culminated in the landmark publication of a guide to Anguilla's reptiles and amphibians and their conservation needs. It is hoped this will stimulate more work to help these threatened and vulnerable species on the islands.

Given the geographical and ecological characteristics of Anguilla, wetlands education continues to be one of the Trust's priorities as it seeks to build national appreciation of the ecological significance of Anguilla's salt ponds and their importance for wildlife. One of the Trust's most successful areas of biodiversity conservation has been its recent work on birds. Recent surveys have shown that the 20 or so private and government-owned wetlands attract thousands of migratory water birds from North America and that the smaller islands such as Sombrero, Scrub, Prickly Pear, and Dog Island hold some of the most important populations of breeding seabirds in the eastern Caribbean. With significant support from the RSPB, the Trust carries out monthly bird counts on the ponds, collaborates with the RSPB on a bird monitoring programme, and has developed the East End Pond Conservation Area as a protected area where great birdwatching can be enjoyed from two specially built observation decks. To strengthen the capacity of the Trust and to provide training for schoolteachers in wetlands education, the Trust has taken part in the Society for the Conservation and Study of Caribbean Birds' West Indian Whistling Duck and Wetlands Conservation project.

ANT heritage tours

While animals play an integral role in the Trust's work, plants have featured prominently as well. The Anguilla Flora project is an ongoing effort being undertaken by the Trust to identify unusual plants and compile a comprehensive list of species found on Anguilla. The list now stands at 500 species thanks to the hard work of a local naturalist and two visiting botanists from the USA and Trinidad who have been identifying and photographing wild flowers and pressing plant specimens for deposit into museum collections. The Trust recently purchased a herbarium specimen case where collections are presently stored for research by scientists, consultants and school groups alike. It is through this project that Anguilla's only endemic plant species *Rondeletia anguillensis* was identified.

One programme that will benefit greatly from the development of this bird guide is the Heritage Tour initiative. These tours are the Trust's contribution to Anguilla's tourism diversification strategy and are designed to show how tourism and the environment can co-exist in a mutually beneficial relationship. The tour is intended to build awareness among local people and visitors of the island's natural and cultural heritage and is an essential introduction to the island beyond the more familiar coastal resorts. The tour includes a history of the eastern portion of the island and stops at the East End Pond Conservation Area, for birdwatching; the Big Spring National Park for a glimpse into Anguilla's rich archaeological history from the Amerindian period; and Fort Hill, with its spectacular views of St Martin. On clear days St Barths, Saba and St Eustatius can be seen and where, with good fortune, occasional sightings of whales are possible at certain times of year.

Drawing on the sentiment of former Trust Council President, Karim Hodge who says, 'Environmental education is not preparation for life – it is life itself,' the Trust has worked with the island's Department of Education to provide an opportunity to enrich the teachings of its educators by including aspects of environmental awareness and education in the primary and secondary school curricula. The Trust produced the *Anguilla sea turtle educator's guide* to support its conservation work and as a creative teaching resource for teachers' use in and outside the classroom. The development of this *Guide to the birds of Anguilla* will serve as a milestone in the missions of both the Trust and the Department of Education. Trust plans to facilitate the use of wetlands as outdoor classrooms by both primary and secondary school teachers and students will be enhanced by the use of the guide for bird identification as well as general and specific information on the wonderful variety of migratory species that flock to Anguilla's wetlands.

The many successful Trust projects already outlined are a testament to the Trust's best endeavours to remain focused on its mission of ensuring the sustainability of Anguilla's national heritage despite the various obstacles, hurdles and shortfalls that it faces. The publication of this bird guide will ensure that the value of birds and biodiversity of Anguilla are given the due attention that they deserve so that the conservation of the ecologically important habitats found across the island and its cays can be secured for future generations.

Anguilla is at a crossroads in its development creating a challenging climate where global influences and local political and economic decisions can have far reaching consequences for the future of the island, its people and its natural resources. In a time of rapid change these external influences provide an equally challenging environment for the Trust to work within particularly as it has little direct influence on land on which the future of so many of the island's historic heritage and wildlife depends. The Trust is well placed to advise and contribute to the essential need to move to a sustainable future for Anguilla. Through a loyal and growing grassroots support and membership it can help build awareness and demonstrate real improvements to nature and culture on the island and to improving the quality of life for residents and visitors. As the Trust responds to the needs of government, which provides its financial viability, and to the genuine concerns that its supporters and stakeholders have over the island's pace of development, the benefits of the Trust's conservation programmes and a high quality environment to the economic and social aspirations of the island will be a key measure of its success.

As we move further into this millennium, the Trust will continue to work towards 'Confirming Our Commitment To Anguilla's Sustainable Development' and achieving 'Preservation for Generations'. The Trust's experience on the road travelled so far suggests a sustainable future will only be achieved through involving people in the processes and by working collaboratively and in partnership. The Trust continues to work with a range of partner organisations within the public, private and voluntary sectors on the island, regionally and internationally. These projects such as the much publicised projects on sea turtles, iguanas and birds have brought in a range of funds from Anguilla and overseas and helped provide regional and international recognition for this work, the island and its natural resources. Such positive action by the Trust can only help Anguilla and its future development.

There is much to be gained from participating in the Trust's work. Volunteering has always been a strong pillar of the organisation. The Trust believes that 'together we aspire, together we achieve and it is only together that we can see Anguilla National Trust as a dynamic people-centred organisation making a positive and lasting contribution to Anguilla's national heritage'. As the Trust moves forward, drawing on the time, efforts and skills of volunteers will be crucial. Please contact the Trust office if you would like to help and contribute to its work. There has never been a more important time to support the Trust and its work. Through its capital

Birdwatching on Road Salt Pond

campaign the Trust is seeking to strengthen its financial base in order to expand the scope of its work and carry out more projects. With the ability for US citizens to make tax-deductible donations to the Trust, there is a joint benefit in supporting the work of the organisation. More importantly, the Trust will need to increase its membership base if it is to survive and if it is to strengthen its legitimacy in the eyes of any detractors but more importantly in the eyes of its partners and beneficiaries. For US$20.00 per year, you can become a member of the Trust that is seeking to ensure that Anguilla's national heritage is preserved for generations of Anguillians and the visitors to its shores. The Trust also produces a range of products to raise money for its work including books, T-shirts, caps and pin badges. If you wish to join or make a donation to the Anguilla National Trust please contact the Executive Director via e-mail at axanat@anguillanet.com or call (264) 497-5297 or visit the organisation's website at www.axanationaltrust.org

Brown booby colony on Dog Island

The Important Bird Areas in Anguilla

Anguilla's biological importance

Conservation International[1] has identified the Caribbean Islands as a biodiversity hotspot, especially for their high number of endemic species, which are found nowhere else in the world. While the hotspot extends over more than 4 million square kilometres of ocean, it covers only about 230,000 square kilometres of land area. In the Caribbean, birds are the group of species where there is relatively the most up to date information on their status. Anguilla is no exception.

Anguilla as part of the Lesser Antilles chain of islands is important for birds both regionally and globally. The network of ponds on the mainland is a key stopover point for the passage of migrant birds flying to and from North America. Its seven small, uninhabited offshore islands are breeding sites for significant seabird populations. It is included in the Lesser Antilles Endemic Bird Area as it shares four restricted-range species with other islands found in the Lesser Antilles chain. These are the green-throated carib, Antillean crested hummingbird, pearly-eyed thrasher and Lesser Antillean bullfinch.

There are three species of global conservation concern recorded on Anguilla. The piping plover occasionally visits during the winter season and the Caribbean coot occasionally breeds at ponds on the mainland. The white-crowned pigeon, once common on Anguilla, was considered locally extinct until a recent sighting in 2005.

BirdLife Caribbean and the Important Bird Area Progamme

The Anguilla National Trust through its collaboration with the RSPB in the UK. is part of a network of conservation organisations from around the world known as BirdLife International, which are working together for people and birds. It is active within the BirdLife Caribbean partnership (www.birdlife.org/regional/caribbean) most recently holding events on Anguilla to celebrate the Caribbean Endemic Bird Festival and taking forward the BirdLife Important Bird Area (IBA) Programme.

1 Conservation International is a US based non-profit organisation whose mission is to 'conserve the Earth's living heritage, our global biodiversity, and to demonstrate that human societies are able to live harmoniously with nature' (www.conservation.org).

Table 1. Criteria for the identification of Important Bird Areas

Category	Criterion
A1 Species of global conservation concern	The site regularly holds significant numbers of a globally threatened species or other species of global conservation concern.
A2 Assemblage or restricted-range species	The site is known or thought to hold a significant component of the restricted-range species whose breeding distributions define an Endemic Bird Area (ie two or more species with a range of less than 50,000 km² which occur together).
A4 Congregations	i The site is known or thought to hold, on a regular basis, >1% of a biogeographic population of a congregatory waterbird species.
	ii The site is known or thought to hold, on a regular basis >1% of the global population of a congregatory seabird or terrestrial species.
	iii The site is known or thought to hold, on a regular basis, >20,000 water birds or >10,000 pairs of seabirds of one or more species.
	iv The site is known or thought to exceed thresholds set for migratory species at bottleneck sites.

The IBA Programme is a global initiative set up by BirdLife International that aims to identify, protect and manage a network of sites which are important for birds using scientifically objective and internationally agreed criteria (see Table 1) so their conservation is secured in the long-term. Important Bird Areas are selected because they hold bird species that are threatened with extinction, have highly restricted distributions or are sites holding significantly large numbers of congregatory species.

The IBA programme began in Europe in the 1980s and resulted in the publication of the *Important Bird Areas in Europe.* It has been extremely successful in securing the formal protection of sites in Europe as the number of designations has increased from 25% to over 60%. This was followed by the publication of IBA Directories for the Middle East, Africa, Asia and the UK Overseas Territories. Directories are under preparation for the Americas and Antarctica. It is estimated that the IBA Programme will identify up to 20,000 sites across the world. This network is considered the minimum needed to ensure the survival of all bird species across their ranges, should habitat be lost elsewhere.

Magnificent frigatebirds

Anguilla's Important Bird Areas

There are four IBAs on Anguilla (see Table 2). The sites cover an area of approximately 594 hectares (ha), which is about 5.8% of the total land area of Anguilla.

The only site on the Anguilla mainland, East End Pond has been selected for species of global conservation concern and restricted-range species. It is a shallow pond (5.25 ha), that occasionally dries out exposing mudflats which are attractive to shore birds. It is now a nature reserve, whose management has been vested to the Anguilla National Trust by the Government of Anguilla.

Table 2 Important Bird Areas of Anguilla

IBA code	Site name	Key species
AI001	East End Pond (Anguilla mainland)	Caribbean coot, green-throated carib
AI002	Dog Island	Sooty tern (12,000–100,000 individuals), brown booby (1,267 pairs)
AI003	Scrub Island	Roseate tern (> 420 individuals)
AI004	Sombrero	Bridled tern (270 pairs)

The other three, Dog, Scrub and Sombrero are all offshore islands and qualify for their populations of seabirds. Sombrero (38 ha) is situated the furthest from Anguilla at 65 km to the north-west. It was previously proposed as a privately developed rocket launch site but the proposal was later withdrawn. It has recently been designated by the Government of Anguilla as a protected area. Dog (207 ha) and Scrub (343 ha) are both privately owned so are at risk of future development.

The IBA process has highlighted gaps in information and the need for more systematic surveys to be taken.

There are four further sites that do not meet the IBA criteria but warrant mention for their regional seabird population significance. These are Anguillita, Little Scrub, Prickly Pear East and Prickly Pear West. The network of ponds on Anguilla mainland may qualify as a global IBA in the future but further surveys are required.

Conservation issues/threats

The main threat to Anguilla's IBAs is uncontrolled infrastructural development. Although draft National Parks and Protected Area legislation has been prepared, at the time of writing it is still waiting to be approved by the Government of Anguilla. Consequently, there are currently no designated National Parks or Protected Areas on Anguilla. As tourism expands and development pressures increase any decisions relating to land use are a challenge as only 5% of land is in government ownership.

At the same time, like many small islands, biodiversity conservation is constrained on Anguilla by limited human and financial resources. Although the Government has increased the resources available to the Environment Department there is a need to invest further in the recruitment, training and retention of staff.

Using the Important Bird Area Directory

It is hoped the IBA sites identified for Anguilla will assist government policy makers and planners in

Sombrero

identifying biodiversity conservation priorities, and progress land development legislation and policies. This is especially important as during the process of compiling the inventory it has become apparent that the sites selected are also often home to other globally important plants and animals.

In addition, Anguilla has signed up to the Convention on Biological Diversity and the Environment Charter, which is a formal agreement with the UK Government to improve environmental management. The management and conservation of the IBAs identified will support Anguilla in meeting requirements under these and other international agreements.

Finally, tourism is the mainstay of the Anguillan economy and its increasing popularity as a tourist destination runs the risk of destroying the natural resource base on which it depends. It is hoped the sites will remain as attractive places for tourists to visit and learn more about the unique wildlife of Anguilla.

How you can help the Anguilla Important Bird Area Programme

The IBA Programme is very much a process. The identification of sites is only the beginning. It is important that action is taken at all the sites to conserve biodiversity and everyone can help. BirdLife have set up the World Birds Database, a system that aims to collect data from local birdwatchers and tourists so that it can inform conservation work and IBA monitoring on Anguilla. If you log-on to www.worldbirds.org and follow the links to the Caribbean, you will find that Anguilla has its own log-in page. In addition to recording your own bird observations, you will be able to view other people's records, which could help you decide where to go on your next birdwatching trip. The checklists can help to keep track of your sightings.

Caul's Pond

Birdwatching on Anguilla

The following sections provide an introduction to birdwatching in the main habitats on Anguilla.

The ponds

The ponds and wetlands on Anguilla attract a range of breeding and visiting water birds and provide some of the best opportunities in the world to enjoy close-up views of migratory shore birds from the far north of the USA and Canada. These birds are particularly active early and late in the day when the light is softer, providing wonderfully atmospheric conditions for birdwatching.

The numbers of birds on individual ponds is dependant on water levels and availability of food. The majority of coastal ponds are seasonal and are replenished by rainfall. It is possible to visit a pond in a particular month one year and find it full of water and teeming with birds and a year later the pond can be completely dry and empty. If there has been a prolonged dry period the number of ponds holding water is much reduced but will include Bad Cox Pond, Caul's Pond and Mead's Bay Pond.

Water birds are adapted in different ways to exploit the food available. Some birds such as grebes and ducks dive to find fish, plants and invertebrates under the surface while others will feed from the surface or reach below the water for food. Herons, egrets and shore birds feed by the edges of ponds or wade in the water. Differing bill and leg length allow each species to take advantage of different prey items. Water levels are important to longer-legged shore birds such as lesser yellowlegs and stilt sandpipers as they will wade in deeper water to find insects and other prey. If water levels are high and large numbers of flies and their larvae are available, these species will gather in large flocks to feed. If ponds are full after heavy rains, short-legged shore birds such as semi-palmated sandpipers and least sandpipers are displaced and have to find ponds with shallower edges.

At times, the ponds can abound with insects and other creatures. Most obvious are brine flies as these can hatch and emerge in large numbers forming clouds on the surface or at the edges of ponds. West End Pond is the best place to see these and to see how different bird species exploit them. Plovers such as Wilson's plover will rush into the swarms of flies and catch them as they fly up. Laughing gulls will wait at the water's edge and snap them up as the breeze blows the insects onto the shore. Other prey includes the adults and larvae of mosquitoes, midges and flies, brine shrimps and a range of worms and other creatures living in the mud. Many of the shallow seasonal ponds do not hold fish, leaving the water birds to feed on the abundance of prey. Without the large flocks of birds there would be many more insects around the ponds and island!

Three birds typify the wetlands and should not be missed. Snowy egrets are the commonest of the herons; elegant with pure white feathers and a long dagger-like bill. They are found near pond edges standing motionless watching for fish and other prey. Sometimes large flocks will wade in the water together. The white-cheeked pintail is the only duck present year round and can often be seen with a party of ducklings in tow. They are often found with black-necked stilts, the commonest breeding water bird. Large

Water birds – West End Pond

flocks of stilts can gather on the salt ponds and their noisy antics, particularly during the breeding season, are a feature of any birdwatching visit.

When you first arrive at a pond it can be a daunting challenge to identify the often wide range of similar-looking species. The larger herons and egrets are the most obvious birds and are often out in the open or perched in fringing trees and shrubs. The great blue heron is the largest and most regularly seen of the grey herons. The white egrets, great and snowy, are easily separated on size, but when alone the larger great egret can be identified by its combination of yellow bill and black legs. Yellow-crowned night-herons and green herons are usually more secretive, although the tiny green heron will fly out into the open to feed. Night-herons are more often active at dusk although with searching you can often find one perched in mangroves or other cover alongside some ponds.

A number of ducks visit Anguilla from North America in the period October–April and most are scarce except for the blue-winged teals. Teals are dabbling ducks and feed at the surface, often in large flocks. The scarcer species are best found by searching through the flocks of blue-winged teals at the small number of ponds they favour.

Small parties of grebes and diving ducks occur more regularly between October and April. The two smallest are the pied-billed grebe, a shy species that usually keeps close to cover, and the ruddy duck which feeds in small flocks out in the open. The only other regular diving ducks are the ring-necked duck and the lesser scaup and both are a little bigger and occur in small parties often at distance on ponds like Caul's Pond although closer views are possible at East End Pond.

Coots and moorhens prefer ponds with weeds and emergent vegetation although both species will walk out onto muddy or grassy edges to feed, where their long toes are visible. Coots will also dive down to pull up plants, bobbing back up in the same area as the dive. Where there are muddy creeks or tree-lined pond edges find a concealed spot and look out for sora rails. These can be difficult to see on the island, but with luck and patience you can see their white-flecked or barred brown and grey plumage and short yellow bill.

The real birdwatching challenge on the ponds is the identification of the wide range of shore or wading birds. Depending on water levels, close views are possible and Anguilla provides a rare opportunity to have shore birds that have flown in from breeding grounds as far away as the high Arctic running around in the Caribbean sun just a few feet away from you! Two tips will help you narrow down the identification challenge, the first is to compare size, shape and bill length and the second is to look at feeding actions. Both of these features can help identification even when the light is poor.

There are few large shore birds found on the ponds. The black-bellied plover is the largest plover with a very short bill. Stilts are distinctive with long legs and a long thin bill. Greater and lesser yellowlegs are among the larger long-legged, thin-billed wading birds both with less distinct grey-brown plumage and bright yellow legs. Lesser yellowlegs is the species that will gather in large flocks on the ponds. They will feed in both shallow water at pond edges or out on the centres of ponds where they swim well. Flocks of lesser yellowlegs will join with stilt sandpipers where their differing feeding actions will help to separate the species at distance.

Rendezvous Bay Pond

Yellowlegs have a more active, walking and picking action while the smaller stilt sandpipers often feed in small tight groups, jabbing into the water in a more concentrated feeding action. Similar but greyer with a heavier bill and pale grey legs is the willet. It is much scarcer than the yellowlegs and obvious in flight with striking white stripes on black wings. The largest wading bird is the whimbrel, a brown bird with a long down-curved bill.

There is a large group of medium- to small-sized wading birds and a further two tiny ones. Of the medium-sized birds look out for the very long bills of the unhelpfully named short-billed dowitcher and the occasional Wilson's snipe. There are four common plover species to check through with snowy plover the smallest and much the paler. Killdeers have two bold breast bands, semi-palmated plover and Wilson's plover one breast band with Wilson's plover having much the heavier bill. The plovers all have a distinctive feeding action and will stand still with their heads held horizontally, looking for prey, before making a short dash to pick an item from the mud. The sandpipers all tend to have a busier, more active, feeding action with head and bills angled downwards. Ruddy turnstones are well-marked with dark upperparts and are mostly found in small flocks, often on rocky shores with the much paler sanderlings. The only other regular species are the stilt sandpiper, small with a long and noticeably drooping bill and long greenish legs, and the spotted sandpiper most often seen singly perched on rocks or feeding on pond edges. Adult spotted sandpipers are spotted below and are easy to identify, but in all plumages their bobbing feeding action and flight just above the water surface, with shallow flicking wing beats and high-pitched ringing calls, is distinctive. If you enjoy the challenge of bird identification, search through the flocks for regular but scarce species such as the red knot, white-rumped sandpiper, pectoral sandpiper and solitary sandpiper. Given Anguilla's location even rarer species can occur such as the ruff and curlew sandpiper from Europe or Asia which were found in 2000.

Solitary sandpiper

Two tiny shore birds are the least sandpiper and the semi-palmated sandpiper. Large flocks feeding out in open water are invariably semi-palmated sandpipers although they will also feed with least sandpipers out on the mud. The flocks of least sandpipers are mostly smaller, and more dispersed as they search for food on muddy shores often well away from water. If you are keen, a few western sandpipers can be found by searching through the feeding flocks of semi-palmated sandpipers.

Whatever your interest, the wetlands on Anguilla remain a rich and vibrant part of the local environment and an hour beside a pond full of water birds can provide a marvellous wildlife spectacle.

Forest and scrub

If you approach Anguilla from the air you look over an island dominated by low, evergreen forest and scrub. The green is broken by houses and other developments and increasingly cut by white limestone tracks. This is a coarse-grained view. Get close to the scrub on land and the wonderful fine grain is evident.

Walk within the low, bushy habitat and you will find much of it is growing on limestone, the trees and plants suited to growth in shallow soils among the rocks. There is a great variety of trees growing here, acacia, licewood, loblolly and thorn trees, with splashes of colour provided by low flowering plants. Large mature trees are few, a result of recent hurricanes and the exposure of the land to prevailing winds and storms. In the more exposed areas the scrub is low, smooth and beautifully sculptured especially in the east

of the island from Mimi Bay to Windward Point and from
Limestone Bay east to Brimegin. In places the scrub is
punctuated by tall spikes of doodle doo cacti and the dark
stems of frangipani plants, adorned with white flowers
after the rains. On the more exposed rocks of Windward
Point is a fantastic field of pope's head cactus.

Turk's head cacti – Windward Point

Land birds are limited on such a small island as Anguilla
but it does mean you can step out in your garden or around
your hotel and quickly see most of the resident species on a
single walk. It is worth exploring this scrubland further and a
few informal trails, well worn by flocks of feral goats, can
lead you across the limestone rocks and into a secluded
world rich in wildlife. There is more local variation in the
plants and insects found across the island but the typical
birds can be seen just about anywhere.

Birds can be seen throughout the day, but are more active in the early to mid-morning period and in late
afternoon. At these times the light is softer and it can be slightly cooler. If you find a good area where birds
feed or take cover, such as a group of bushes, stand in a shady spot and you may see birds such as warblers
move through and enjoy close views of birds feeding.

Access is easiest in the areas of limestone plateau where the rock is at the surface. Many of the
bushes have thorns so you need to take care. You can usually find some open paths to explore but be
sensitive over access as much of the island is in local, private ownership. The scrub vegetation comprises
a wide variety of shrubs and low bushes in a range of shapes and most are evergreen, their fleshy or

Mimi Bay Pond

shiny leaves are adapted to conserve water in the
dry conditions.

The scrub is generally a quiet world except for
the sharp contact calls of small birds or the soft song
of Zenaida doves. Walk a few steps however and the
peace is quickly broken as you disturb a large ground
lizard. These land lizards make a scurrying retreat
made all the louder as they run across dried leaves
that gather under bushes. One of the first birds you
are likely to see is the tropical-looking bananaquit
with its flash of sulphur yellow on its lower breast.
Noisy and active, they move through the bushes,
often gathering in small groups where trees are in
flower. You will get to know them well as they are the
commonest land bird on Anguilla and are also found
across the Caribbean. Bananaquit nests are the most
often seen too – an untidy ball of grasses suspended on the outer branches of bushes.

The only other small, bright yellow bird is the resident yellow warbler. A sharp *tzip* call is the first sign
that one is nearby. Yellow warblers are common year round but from September to April you may be lucky
and come across one of the colourful North American warblers that pass through on migration or spend the
Northern winter here. The variety and numbers of warblers varies annually and may be linked to stormy
weather systems in September and October forcing over-flying migrating birds to land.

One of the great thrills of tropical forests and gardens is the sight of a colourful hummingbird. They
often appear at speed without warning except for a loud whirr of wings. They can be difficult to watch as
they move from flower to flower feeding using a long curved bill. With patience you may see them fly to a
favourite perch. The only common species is the green-throated carib. It is a brilliant iridescent mid-green
and the males have a vivid purple breast patch.

The most common bird of prey on the island is the American kestrel. It is also the only bird of prey that
breeds and is likely to be present in the period May to August. It searches for prey from a high perch and will
fly down to take lizards and large insects, flying onto a branch or post to eat.

There are two resident finches on the island and both can be difficult to find at times. Black-faced grassquits are the commonest, are very vocal, and will often gather in small flocks in grassy areas. In the heat of the day grassquits will repeatedly visit a good source of fresh water. It is worth searching for the Lesser Antillean bullfinch although it can be surprisingly retiring. The mature gardens of some of the hotels can be among the easiest places to see it although it is a widespread species on the island. They will feed unobtrusively on trees buds and fruits often betraying their presence with a short trilling call.

If you get the chance, a walk in the taller woodland is a different experience. The canopy is often closed, the air more humid and there is a wonderful variety of leaf shapes and patterns of light and shade. Bromeliads grow on some of the trees and hermit crabs in their shells make a soft thud as they fall to the ground from branches. Birdwatching can be difficult in the densely wooded areas, but at sites like the Katouche Valley, North American warblers join yellow warblers, bananaquits and hummingbirds. These areas provide a tropical forest environment where the defining sound comes from the pearly-eyed thrasher with its range of single whistles, squeals and short phrases cutting through the still damp air. These small forest fragments still offer a window into the rich and complex world of the tropical forest.

Shoal Bay East

The beach

Anguilla's beautiful white beaches curving around shallow, sandy bays and enclosing clear, warm seas provide an idyllic retreat for residents and visitors. They are at their busiest from November through to April as tourists escape from the colder north and the pressures of modern living to wind down and relax. If you take time to look up from the holiday reading and look out across the bay you can enjoy close views of some of the most spectacular seabirds in the Caribbean.

Seabirds do not come much bigger than brown pelicans. They look heavy and ungainly when perched on a moored boat or sitting in small groups on a rocky headland. A few heavy wing beats later and they can soar or glide with ease, often flying along the coast in formation, before suddenly breaking their flight and twisting to make a dramatic, near vertical dive into the sea to catch fish. Look out for the way they streamline their large bill and body with wings angled tightly backwards in one of the most remarkable feeding actions of any large bird.

Most of the seabirds searching for fish in the shallow seas around Anguilla have perfected some form of dive to surprise and catch prey. These include the brown booby, the royal tern, and a few red-billed tropicbirds. These birds are present year round and from late April until September they are joined by laughing gulls, and Sandwich, roseate and least terns. These seabirds are very easy to see from most of the island's beaches and if you walk out or swim in the sandy bays or inshore coral reefs you will quickly see why. The seas hold a rich variety of fish species, mostly pale coloured over the sand and more cryptic or more colourful within the reefs.

Seabirds will congregate where fish gather in shoals or are pursued by larger predatory fish. You can often see these shoals breaking the surface and streaking the sea with flashes of silver. Suddenly a group of pale grey birds gather and begin to dive, making small splashes as they catch prey near the surface. These are terns, their white underparts invisible to their prey below. A few larger brown birds move in low over the water flying with a few stiff flaps, long wings slightly arched then a glide at a slight

Immature brown boobies

angle with one wing tip almost cutting the surface. These are brown boobies and they will dive from close to the surface at a narrow angle or lift to dive near-vertically becoming fully submerged. Brown boobies breed on Anguilla's offshore cays and both dark brown adults with a striking white belly and pale brown young birds can be present.

In the period April–September, laughing gulls with their dark heads, smoky grey wings with black wing-tips dance in attendance on the true fish-catchers hoping to gather scraps or steal prey. In Mead's Bay a laughing gull has been seen to land on the back of a brown pelican and try to steal a fish from its bill! The action at these feeding 'frenzies' can end as quickly as it starts as seabirds drift off leaving a few gulls sitting on the surface.

Look up from feeding seabirds and you may well see a large, dark bird hanging high in the air. These are magnificent frigatebirds, one of the iconic birds of the Caribbean islands. From effortless drifting on long pointed wings, with long forked tails, they suddenly power into action in pursuit of the terns, twisting and turning in close pursuit as these pirates of the air try to force the terns to drop their prey. It is yet another moment of high drama played out every day over the beaches and inshore waters of Anguilla.

The smaller islands and cays

Take a boat out from the mainland and it will open up a whole new perspective on the islands. Away from the beaches, much of the coastline is low limestone or sandstone cliffs. These often-isolated coastal areas provide roosting areas for brown pelicans and brown boobies and in the period October–March belted kingfishers will dive into the surf. The coast is increasingly sought after for residential development and the wild and rugged nature of the rocky coastline is rapidly changing leaving less room for space, solitude and wildlife.

Head out on a boat to the cays and you get a glimpse of how the first settlers will have come across the islands of the Lesser Antilles. Islands such as Dog and the Prickly Pears emerge from the beautiful blue

Prickly Pear West

expanse of sea and brilliant white wave caps. Low cliffs and white beaches come into sharper focus behind clear inshore waters and coral reefs. The initial approach can be surprisingly bird free as many seabirds avoid the mainland and the stretch of sea in between, venturing far out to sea in search of fish and other marine prey. Once you reach the cays the contrast is immediate as tens of thousands of seabirds seek out these remote uninhabited islands and cays to breed, forming large and noisy seabird cities.

The breeding colonies are full of activity. Birds will go through courtship rituals at nest sites and adult terns will wheel over the islands in noisy display flights. As chicks appear, the adults increasingly fly to and from feeding grounds providing them with food. They often run the gauntlet of waiting frigatebirds. An approach by boat is a great way to see these remarkable birds. Most inquisitive are the brown boobies, especially the milky brown immature birds that will hang in the air above the boat. Around Sombrero, Dog Island and Prickly Pear West you will also see the white and black masked booby, a scarce marine bird that is most easily seen around its breeding islands.

The offshore cays vary in size from small areas of rock just a few acres in extent to Scrub Island, which covers 850 acres (343 ha). They are mostly similar to the mainland with a mix of low cliffs, beaches and coastal lagoons. Water birds, typical of the mainland, are found in low numbers on the lagoons as are land birds in the scrub and although fewer species can be found, peregrines are more conspicuous and haunt the larger islands. Remote Sombrero is famous for attracting rare migrant land birds from North America.

The vegetation is typically shorter on the more exposed cays and is largely absent from the rocky outcrop of Sombrero. These low islands are mostly arid with little fresh water and the dense low prickly scrub is subject to constant winds and salt spray from waves breaking on rocks. The cays are best left undisturbed to allow seabirds to breed but Prickly Pear East is accessible for visitors and here you can see the low thorn bushes and prickly pear cacti that make an impenetrable barrier to the centre of the islands while providing a safe place for seabirds to nest. This harsh, inhospitable environment is perfect for seabirds and marine turtles to go through their breeding cycles and the cays of Anguilla are internationally important for both.

A trip around the coast of Dog Island and Prickly Pear West will provide views of all the seabirds. Boobies nest on a narrow bare strip around parts of the cays and you can see both adults and chicks in the prolonged breeding season that runs from November to August and may be year round in response to food availability. Young booby chicks are white and downy and often sit upright on the cliff tops as the adults leave them to find food. A highlight of any visit is a view of a tropicbird. Their gleaming white plumage and long tail streamers bending in the sea breeze is a strikingly beautiful image of these tropical islands.

The smaller terns arrive from April onwards and in some years huge numbers seek out these undisturbed islands. The black and white sooty tern is the commonest bird on Anguilla and up to 100,000 can be found on Dog Island alone. These are joined by the very similar bridled tern and the brown noddy; a beautiful chocolate-brown seabird that nests on the cliffs. Together they symbolise the high quality marine environment around the islands and the importance of the safe nesting sites that the islands provide as all these terns spend the majority of their lives far out to sea.

Red-billed tropicbird

Several species of grey tern are found on Anguilla and these prefer flat sandy or rocky locations to breed such as Scrub Island and the small island of Anguillitta. The largest of these terns, the royal tern, can be found year round with the other smaller terns present from April to August with a few birds lingering into November. If you are unable to visit the cays several of these birds will visit the more remote coasts and ponds on the mainland including Windward Point and the cliffs at West End. Often small groups of Sandwich and roseate terns will join royal terns to rest on beaches or salt pond walls.

Whether you are fortunate to sail around a remote island such as Dog Island or catch views of seabirds in the inshore waters of the mainland, you will be witnessing part of one of the great seabird populations of the Caribbean.

Identifying 60 common species

How to use this section

With a little effort it is possible to see 50–60 species of bird in Anguilla on the shortest of visits. This section will help you identify them.

Species groupings

This guide is primarily aimed at beginners to bird identification and for ease of use we have grouped birds by the three main habitat types they are found in: sea and coast (**seabirds**); ponds and wetlands (**water birds**) and forest and scrub (**land birds**). Some species such as the brown pelican typically feed along the coast but will also visit ponds to feed or rest. In these instances we have placed the species in its more typical habitat to avoid duplication. Similarly some warblers will feed within trees and bushes that are found around ponds and wetlands. To enable quick comparison of species we have included these birds of the wetland edge within the section on land birds.

Choice of the 60 species

All the species illustrated in this section are common or occur regularly on the islands. Many can be found year round with little effort, others can usually be found by visiting the right habitat at the right time of year. It has proved difficult to decide on the final choice and species that nearly made it onto this list are: white-tailed tropicbird, tri-coloured heron, lesser scaup, peregrine, merlin, sora rail, American golden plover, red knot, pectoral sandpiper, Wilson's snipe, white-winged dove, Antillean nighthawk, Antillean crested hummingbird, black-and-white warbler and house sparrow.

Bird names and order

The first name given is the general common name each species is known by and for these we have followed the *Birds of the West Indies* by Rafaelle *et al*. As some birds have a wide global distribution, common names may vary and most identification guides refer back to the scientific name given to each species as part of a wider classification of the natural world. We have again followed the systematic order and scientific names set out in *Birds of the West Indies*. Many common and familiar species have been given a local name. For example Anguillians know the common Zenaida dove as the turtle dove. In the Caribbean these local names often vary from island to island. In conversations on the island it appears that some local names are used more than others and we have included the more widely used local names in this guide.

How to identify species

This section includes the main identification features of each species and is supported by a colour photograph. To give an idea of size we have given the average length of the bird (**L**) and where it is helpful for identification the wing span (**WS**). As some species vary in size or males and females differ in size, we have given an average length.

Some species also occur in a range of different plumages from initial juvenile plumage to adult plumage and in some species the male and female have different plumages. In this guide we have been unable to describe or illustrate each plumage variation but where appropriate we have included information on the most typical plumages seen on Anguilla. For example all shore birds have been photographed in their non-breeding plumage.

Similar species

We have only included text in this section where there is a chance a species might be mistaken for other similar species. Again we have only added species here in context with the species likely to be seen on Anguilla. In other parts of their range there may be other species that are very similar but if they are rare or unlikely to be seen on the island they have not been included. If a species you have found does not look quite right please take notes and a photograph if possible and inform the staff at the ANT who may be able to help.

When to see it

This section includes the time of year species are typically seen. Resident birds are present year round and many migratory birds will spend several months on the islands. For more detail refer to the monthly checklist. We have also tried to give an idea of the status of each species in this section including information of how often it is seen and typical numbers present.

Where to see it

Much of the data from recent bird surveys is collected by site and if you enter your records via the internet database **www.worldbirds.org** you will also be able to record your sightings for each site you visit and even create new sites such as your garden. From the recent information we have identified the sites where in recent years species are most likely to be present. For example if you visit Bad Cox Pond or East End Pond in December you are very likely to see blue-winged teals. The only cautionary note is with wetland birds. The site information for seasonal ponds is based on observations when the pond holds water. Those seasonal ponds that dry out completely at times will support very few birds when dry.

This section can help narrow down some identification challenges. For example in the period December–March the only grey tern likely to be present along the coast or on salt ponds is the royal tern as it is the only resident tern.

Local notes/notes

We have added a few observations and information of more local interest in this section as an encouragement for you to make your own notes!

Other birds

These sections at the end of each group of birds include all other species we have records for in the period 1990–2006. It includes information on status, favoured sites and specific details of scarce or rare birds seen on Anguilla. The identification of several of these species is covered in the **Similar species** section of some of the common 60 species.

SEABIRDS

Red-billed tropicbird
Phaethon aethereus

HOW TO IDENTIFY IT L 48 cm (19″), WS 110 cm (44″). The length does not include the tail plumes. A striking, all white, seabird with black wingtips, a dark line through the eye, greyish barred back, red bill and distinctive very long tail streamers. Tropicbirds fly along the coast where they can be seen diving vertically from height after prey or flying into nest holes on cliffs where birds will often make harsh rattling calls.

SIMILAR SPECIES The scarcer white-tailed tropicbird also nests on the island and is very similar except for black diagonal lines on the inner part of the upper wing, a plain white back and a shorter yellow bill.

WHEN TO SEE IT A few can be seen year round although it is more easily seen during the main breeding season from January to June. The population is currently fewer than 25 pairs.

WHERE TO SEE IT The cliffs at Little Bay are the most reliable site for this species and here you can also find the scarcer white-tailed tropicbird. Away from the breeding season, they are occasionally seen around rocky headlands such as Windward Point. A few pairs also nest on the offshore cays.

LOCAL NOTES Tropicbirds tend to fly into the cliffs at set times of day and are then absent for long periods.

YOUR NOTES

Masked booby
Sula dactylatra

LOCAL NAME
White booby

HOW TO IDENTIFY IT L 85 cm (34") WS 155 cm (62"). A large and striking seabird, it is all white except for its dark face in front of the eye, black flight feathers forming a trailing edge to its wing and a black tail. The bill is pale yellow at all ages and the feet are grey. Immature birds are darker and have a dark head, wings, back and tail with a white collar and rump. The downy chicks of boobies are white until the first feathers appear.

SIMILAR SPECIES Adult brown booby is all dark brown above and has dark head and breast. Immature masked booby is closer in plumage to adult brown booby but has a white fore-neck, collar and rump and lacks the bright yellow feet of brown booby. Red-footed booby adults occur in a white and brown phase and have bright red feet and a white tail.

WHEN TO SEE IT A few are present year round with the main breeding season between December and August. The current population is less than 100 pairs.

WHERE TO SEE IT These birds are only found on the rocky plateau and cliff tops of Dog Island and Sombrero. A boat trip around Dog Island presents the best opportunity to see one, as there are no records around the coast of mainland Anguilla.

LOCAL NOTES They will lay one to two eggs in a shallow scrape but invariably will only raise one chick. Masked booby is a rare bird in the region and Anguilla may hold 10% of the Caribbean population.

YOUR NOTES

Brown booby
Sula leucogaster

HOW TO IDENTIFY IT L 74 cm (29"), WS 140 cm (57"). A large, slim seabird, it has a distinctive shape with long, pointed wings and a long neck with a pointed bill. Adults are dark brown with a clean white belly and undertail and have a yellow bill, yellow legs and feet. Immature birds are a paler milky brown with a grey bill. The underwing has a pale stripe along the inner part in all ages. They fly low over the water with stiff wing beats and glide with one wing tip almost cutting the surface of the sea.

SIMILAR SPECIES See text under masked booby. Both masked and red-footed boobies are found on offshore cays.

WHEN TO SEE IT It is present year round on the coast with breeding colonies on Dog Island, Prickly Pear East and West and Sombrero. The main breeding season is January to June however this species has been known to nest at all times of the year. It is the commonest of the three booby species with a population in excess of 2,000 pairs in recent years.

WHERE TO SEE IT Away from the breeding colonies brown boobies can occur anywhere where fish are shoaling in inshore waters but especially at Crocus Bay, Mead's Bay, Little Bay, Shoal Bay East, Island Harbour and West End.

LOCAL NOTES It often joins other feeding seabirds by flying in low over the water in search of fish before making shallow dives. Look out for them resting on small moored boats or jetties as at Scilly Cay or on the cliffs at Little Bay.

YOUR NOTES

Brown pelican
Pelecanus occidentalis

HOW TO IDENTIFY IT L 128 cm (51"), WS 198 cm (79"). The largest seabird with a distinctive shape: large bill with pouch, heavy body and short tail. Adult birds have white on the head and neck, red-brown on the rear neck and dark underparts. Immature birds are a plainer, greyer brown, are white underneath and have grey bills. The plumage differences make it easy to separate adults and young birds when they fly past together. Pelicans are well known for making spectacular plunge dives to catch fish in inshore waters. They often fly in small lines with slow flaps and glides on long, broad wings.

WHEN TO SEE IT Pelicans can be seen year round with the largest numbers between October and April. They are most often seen in small flocks but they will congregate in larger parties of 30 or more to fish. A few pairs breed occasionally on the outer cays.

WHERE TO SEE IT A familiar sight around the coast, they can often be seen sitting on rocky headlands or small boats moored offshore. They occur in larger numbers from Crocus Bay to Limestone Bay, Shoal Bay East to Island Harbour and around West End. Small numbers are often seen over ponds or resting on salt pond walls with Caul's Pond and Mead's Bay Pond among ponds that also attract birds to fish. Small groups will occasionally fly across the mainland.

LOCAL NOTES Pelicans often surprise visitors to Little Bay by diving into the sea just a few metres away!

YOUR NOTES

Magnificent frigatebird
Fregata magnificens

HOW TO IDENTIFY IT L 98 cm (39"), WS 225 cm (90"). A large distinctive seabird with a deeply forked tail and long, thin, angular wings which are pushed forward at the bend. It can look all dark from a distance but when closer the males are all black, females have a white breast and immature birds have a white breast and head. A pirate of the seas, it will dive and pursue terns and other seabirds forcing them to drop or disgorge prey. They also pick up food items from the sea but never alight on the water. Males have a bright red throat patch that is inflated during its breeding display.

WHEN TO SEE IT They can be seen year round with immature birds more frequent following the breeding season. There is a small breeding colony on Dog Island where the breeding season appears to run from December to May.

WHERE TO SEE IT Frigatebirds can be found anywhere above the coast especially near headlands. Favoured areas are around Sandy Ground, Shoal Bay East, Island Harbour and Scilly Cay and Windward Point.

LOCAL NOTES This is one of the defining birds of the Tropics and Caribbean and is a constant feature of the coast as they hang high on the breeze. Look for them in the sky above groups of feeding terns and other seabirds.

YOUR NOTES

Laughing gull
Larus atricilla

Laughing bird

HOW TO IDENTIFY IT L 41 cm (16.5"), WS 100 cm (40"). This small gull with a black hood has a smart plumage of smoky grey wings with a white rear edge and black wingtips contrasting with white underparts and tail. The bill is red in the breeding season and appears drooped at the tip and it has black legs. In flight, you can see black tips to the white underwings. It is vocal in flight giving a short harsh laughing call. Young birds can be confusing and are brown with a black band near the tip of the white tail.

SIMILAR SPECIES The only gull regularly found on the island.

WHEN TO SEE IT This is a common breeding species visiting mainly in the period April–October with a few, mostly immature birds, seen in other months. Breeding colonies are found on Dog Island, Prickly Pear East, Scrub Island and Sombrero. The main arrival occurs in April when large numbers of adults, often in excess of 300 birds, gather on salt pond walls at Road Salt Pond.

WHERE TO SEE IT A familiar sight across the island during the months it is breeding and can be seen almost anywhere although most sightings are along the coast including Mead's Bay and Shoal Bay East. They will visit some ponds in large numbers when food is abundant such as at Road Salt Pond and West End Pond. Young birds from breeding colonies on the cays begin to arrive on the mainland in August.

LOCAL NOTES Although it is a common bird on Anguilla, there are very few records outside the breeding season and many visitors to the island never see a gull. Laughing gulls will land on the backs of pelicans ready to pick up any scraps.

YOUR NOTES

Royal tern
Sterna maxima

HOW TO IDENTIFY IT L 50 cm (20"), WS 102 cm, (42"). A large tern looking all white at distance but with pale grey upperwings, blackish wingtips, a shallow forked tail, black cap to the eye and a sturdy orange bill. Away from the breeding season the cap is often reduced to a crescent from behind the eye around to the rear crown. Young birds have more black on the upper wingtips and a black line across the front of the inner wing. Terns fly over the sea on the lookout for fish, diving in from height to catch prey. When fishing, they are often the targets of frigatebirds, which chase them in spectacular aerial pursuits in an attempt to force them to drop their catch.

SIMILAR SPECIES No other terns or gulls are around the coast in the period December–March except for the occasional rare or unseasonal visitor. From April to October, Sandwich terns are present and are slightly smaller and paler with a thinner black bill with a pale tip. Roseate terns and common terns are even smaller and have much smaller and narrower bills.

WHEN TO SEE IT Royal terns are present year round around all coasts and breed on a few of the cays with up to 100 pairs present. Feeding or roosting flocks are mostly fewer than 10 birds.

WHERE TO SEE IT You can find terns fishing in inshore waters anywhere around the coast but they will congregate where fish are located. Mead's Bay, Shoal Bay East and Island Harbour all attract terns and some gather on secluded stretches of beach, rest on buoys or small moored boats. They often fly into ponds to roost on salt pond walls particularly on Road Salt Pond, Long Pond and West End Pond.

LOCAL NOTES One of a handful of resident seabirds typical of the islands.

YOUR NOTES

Sandwich tern
Sterna sandvicensis

HOW TO IDENTIFY IT L 42 cm (17") WS 85 cm (34"). A very pale tern, it appears white at distance. The wings are a pale grey and they have white underparts, an untidy black cap and a long thin black bill with a yellowish tip. The tail has a short fork to the tip and Sandwich terns look short-tailed in flight. Non-breeding birds have a white forehead to above the eye.

SIMILAR SPECIES See text under royal tern.

WHEN TO SEE IT Visits the offshore cays of Anguillita and Scrub Island with 100 to 200 pairs arriving to breed in April. Most birds leave by September with a few birds remaining to fish inshore waters beyond this. Small numbers visit the coast and ponds with flocks typically of up to 25 birds.

WHERE TO SEE IT Look for it in flocks of feeding terns and other seabirds in inshore waters such as in Crocus Bay, Mead's Bay and Shoal Bay East. They also follow royal terns onto some ponds to rest on sand banks or salt pond walls with Long Pond, Road Salt Pond and West End Pond regularly attracting birds.

LOCAL NOTES The third commonest tern seen on the mainland in the breeding season and middle-sized between the other two, royal tern and least tern.

YOUR NOTES

Roseate tern
Sterna dougallii

HOW TO IDENTIFY IT L 38 cm (15"), WS 72 cm (29"). This is a medium-sized tern and is very pale grey above with a whitish panel across the rear edge of the upper wing. It has long tail streamers, a dark cap and a narrow dark bill with some dark red at the base.

SIMILAR SPECIES See text under royal tern. A few common terns pass through the islands and may breed. The two species are very similar although adult common terns have a dark tipped brighter red bill and a small black wedge on the rear edge of the wingtips. A field guide should be used to identify the smaller grey terns in non-breeding and juvenile plumages.

WHEN TO SEE IT Arrives to breed on Scrub Island in April and most leave by September. Recent counts suggest 100–200 pairs breed. Small numbers can be seen off the mainland coast from May to September.

WHERE TO SEE IT Can be seen without visiting the breeding colonies with a few birds regularly flying close to the beach off Shoal Bay East and Windward Point. A few will occasionally join resting terns on beaches, salt pond walls and sand banks.

LOCAL NOTES A scarce and declining bird worldwide, it formerly bred on Sombrero. A new local breeding colony was only recently discovered.

YOUR NOTES

Least tern
Sterna antillarum

HOW TO IDENTIFY IT   L 22 cm (9"), WS 50 cm (20"). This is a very small tern, white except for pale grey upperwings and a black stripe down the front of the outer wing. Least terns have a dark cap with a white forehead. The bill is fine and yellow with a dark tip in breeding plumage becoming black as the season progresses. They have long thin wings and a short, slightly forked tail. Often vocal with shrill high-pitched calls, they can be difficult to see when high overhead. They hover more than the other terns moving location regularly before making shallow dives to catch fish.

SIMILAR SPECIES   Very small size and bill colour with pale forehead should separate from all other small terns in the breeding season.

WHEN TO SEE IT   Arrives in April and leaves in September, it is the only tern breeding on the mainland and a few nest on the outer cays with a recent total population of 200–300 pairs. This species is susceptible to disturbance and former colonies at Cove Pond, Mead's Bay and West End Pond have been lost since tourism increased. A few, mostly immature birds, can be found into October.

WHERE TO SEE IT   This is a widespread bird during the breeding season visiting coastal areas and ponds. It nests in several locations and care should be taken not to disturb breeding colonies. Typical sites to see feeding birds are Road Salt Pond, West End Pond, Cove Bay Pond and Long Pond.

LOCAL NOTES   Will sometimes gather away from water with flocks often seen resting on the runways at Wallblake airport. An occasional first summer bird has been seen with the adults.

YOUR NOTES

Sooty tern
Sterna fuscata

Egg bird

HOW TO IDENTIFY IT   L 40 cm (16"), WS 80 cm (32"). This is a striking tern, black above and white below with long, narrow wings and deeply forked tail. Look out for white outer tail feathers, white forehead only reaching the eye, and dark bill and legs. These birds have a three-note *wide a-wake* call.

SIMILAR SPECIES   The bridled tern also visits the offshore islands to breed and the two species can be difficult to separate when seen at sea. Bridled tern has slightly paler dark grey-brown upperparts and central tail. The black cap is similar but with good views, the narrow white forehead patch extends above and just beyond the eye.

WHEN TO SEE IT   These birds can only be seen in the breeding season from April to September. Numbers can fluctuate annually with recent estimates of the large colony on Dog Island varying between 10,000–50,000 pairs.

WHERE TO SEE IT   Another marine tern, the easiest way to see it is by boat on a sail or trip around Dog Island with small colonies also on Scrub Island and Sombrero. Very occasionally, a single bird has wandered into the inshore waters around the mainland of Anguilla.

LOCAL NOTES   Although many locals and visitors never see this bird, it is by far the commonest species in the islands. Safe nesting islands and an availability of food make Anguilla highly important for this and other species of seabird within the Caribbean.

YOUR NOTES

Brown noddy
Anous stolidus

HOW TO IDENTIFY IT L 38 cm (15"), WS 80 cm (32"). A distinctive tern, this species has a chocolate-brown plumage except for a white forehead that merges to grey on the crown. The flight is buoyant and it has a short, squarer-tipped tail. Young birds lack the white forehead and pale crown.

SIMILAR SPECIES All brown underparts separate from all terns except for juvenile sooty terns, which differ in having white underwings and white-spotted upperwings and back.

WHEN TO SEE IT Arrives with other terns in April to breed on offshore cays and mostly leaves by September. The population of 1,000–2,000 pairs breeds on five of the cays.

WHERE TO SEE IT A cliff-nesting bird, it is easier to see from boats around Dog Island and Prickly Pear East. A few have been seen from the mainland recently along the cliffs at the extreme West End.

LOCAL NOTES This species will nest in trees but the Anguilla population appears only to nest in rocky areas or on ledges and crevices on cliffs.

YOUR NOTES

Other seabirds

White-tailed tropicbird
Phaethon lepturus
A rare breeding visitor first reported in 1999. A few breeding pairs have been seen from March–July at Little Bay on the mainland and at Prickly Pear East.

Red-footed booby
Sula sula
A rare occasional breeding visitor to the offshore cays with one to two pairs nesting in low bushes on the offshore cays. The last confirmed breeding was in 2000.

Ring-billed gull
Larus delawarensis
A rare visitor, immature birds were seen at West End Pond in 1996 and over Dog Island in 2000.

Lesser black-backed gull
Larus fuscus
A rare visitor, two juvenile birds were present at several sites on the mainland in 2006.

Great black-backed gull
Larus marinus
A rare visitor increasing in numbers with first year birds at Road Salt Pond in 2001 and off Blowing Point Pond in 2002 and an adult at Scilly Cay in 2005.

Caspian tern
Sterna caspia
A rare visitor with just one recent record of two birds on Dog Island Pond in May 2004.

Common tern
Sterna hirundo
A scarce visitor and possibly a breeding visitor to the cays seen singly or occasionally in parties of up to 20 birds in the period April–November. Most records involve birds joining other species of terns resting on beaches and salt ponds or flying offshore.

Bridled tern
Sterna anaethetus
A breeding visitor, up to 500 pairs are present from April–August on several cays with the largest population on Sombrero. Anguilla holds up to 10% of the Caribbean breeding population of this species.

Black tern
Chlidonias niger
A rare visitor more typically seen over wetlands as with the only record of two birds at West End Pond in 1999.

WATER BIRDS

Pied-billed grebe
Podilymbus podiceps

HOW TO IDENTIFY IT L 32 cm (13"). A small, compact water bird, it is grey-brown with tawny flanks, has a short thick bill and no obvious tail. Adults in breeding plumage have black face markings and a black band around the bill. It feeds by diving and can be elusive and hard to re-locate after a dive.

SIMILAR SPECIES The only comparable species are diving ducks: ring-necked duck, lesser scaup and ruddy duck. Ruddy duck is nearest in size and the browner female has a longer grey bill, a dark crown and line through its eye and an obvious spiky tail.

WHEN TO SEE IT These are mainly non-breeding visitors from North America in very small numbers mainly from October to April. It is usually seen singly or in small groups of two to four.

WHERE TO SEE IT The freshwater ponds at East End Pond and Caul's Pond are favoured and these provide sheltered areas and vegetation for cover. Occasionally visits other ponds.

LOCAL NOTES A large party of 17 adults and immature birds at Caul's Pond in June 2000 suggested breeding had taken place.

YOUR NOTES

Great blue heron
Ardea herodias

Large gaulin

HOW TO IDENTIFY IT L 115 cm (46"). This is a large, tall, grey wading bird with a long bill, black stripe behind the eye, a paler neck and grey, not black, legs. These are the largest regular wading birds and are bigger and heavier than great egrets. Immature birds are dirtier grey with a black crown and a more streaked, grey neck. They are usually found singly resting by ponds or wading in deep water. These herons can be local and wary and often give a harsh *fraank* call as they fly off.

SIMILAR SPECIES The only species likely to be confused with great blue heron are the smaller and much scarcer tri-coloured and little blue herons. Tri-coloured herons have a distinctive clean white belly and undertail contrasting with a dark grey or brown neck and breast. They also have white under the wing and are long-billed. Little blue herons are much smaller and all dark grey with paler legs. See a field guide for plumage variations in these species.

WHEN TO SEE IT It is a regular non-breeding visitor almost exclusively seen in the period October–April. Often seen singly but occasionally arrives in small flocks when migrating with parties of over 10 birds exceptional.

WHERE TO SEE IT For such a large bird they can be elusive but the large freshwater ponds such as Caul's Pond, East End Pond and Bad Cox Pond usually hold a few birds but they are mobile and can turn up on any wetland including small ponds such as Katouche Bay Pond. .

LOCAL NOTES On large ponds they often space themselves out and several can be seen sentinel-like along the edge as they wade slowly in still waters waiting for fish or other prey.

YOUR NOTES

Great egret
Ardea alba

LOCAL NAME
White gaulin

HOW TO IDENTIFY IT L 98 cm (39"). A tall all-white water bird it has a long slender neck, yellow bill and all dark legs and feet. It is often seen singly wading in ponds or perched in low trees. When seen with a mixed group of herons and egrets it is easily picked out on size from the smaller snowy egret. A narrow line from the base of the bill under and behind the eye is distinctive.

SIMILAR SPECIES See text under snowy egret.

WHEN TO SEE IT A few are present year round but essentially, it is a non-breeding species with more birds around in the September–April period. Usually fewer than five birds are seen together at any single site.

WHERE TO SEE IT East End Pond, Caul's Pond, Little Harbour Ponds and Road Salt Pond are favoured but it can turn up on any wetland area. Occasionally one is seen away from wetlands looking for insects and lizards in grassy areas and among sand dunes.

LOCAL NOTES Numbers appear to be slowly increasing on the islands.

YOUR NOTES

Snowy egret
Egretta thula

LOCAL NAME
Gaulin

HOW TO IDENTIFY IT L 60 cm (24"). A striking all-white water bird, snowy egrets have a long dark bill and yellow in front of the eye. The dark legs contrast with bright yellow feet that are especially obvious in flight. Immature and non-breeding birds have varying amounts of yellow on the legs. This species grows fine showy plumes used in its display. They are invariably associated with water where they can be found stalking the shallow edges of ponds for fish and other prey.

SIMILAR SPECIES Great egret is larger with a longer, slender neck and a distinctive combination of long yellow bill and all dark legs and feet. Cattle egret is smaller, stockier with shorter dark legs and a shorter yellow bill. All three species will roost together.

WHEN TO SEE IT A few are present year round with numbers boosted by an influx of birds from North America from October to April. They are by far the commonest egret on the island although typically fewer than 20 birds will be present on individual ponds with flocks of 100 to 150 exceptional. There is no evidence of breeding to date.

WHERE TO SEE IT Typically found across the island they can be seen on any pond although they favour ponds with higher water levels where they will occasionally gather in large numbers. Caul's Pond, Blowing Point Pond, East End Pond and Little Harbour Ponds attract the largest flocks. Egrets will gather in late afternoon to roost in bushes by ponds such as Road Salt Pond and Caul's Pond.

LOCAL NOTES As seasonal ponds dry out spectacular feeding flocks can gather in the shallow waters such as at Caul's Pond.

YOUR NOTES

Cattle egret
Bubulcus ibis

Cattle gaulin

HOW TO IDENTIFY IT L 50 cm (20"). This is the smallest of the white egrets with a short yellow bill, pale eye, short stocky neck and dark legs. It is more often seen away from wetlands feeding in agricultural areas or around cattle, sheep and goats in grassy areas where it picks up any large insects they disturb.

SIMILAR SPECIES See text under snowy egret.

WHEN TO SEE IT Small numbers of this egret are present year round but it remains a non-breeding species with over 100 birds present in the September–April period.

WHERE TO SEE IT It favours dryer areas to feed on farmland and grasslands including cricket pitches and the golf course. The easiest place to see them is around the agricultural area in the centre of The Valley. A large flock of around 80 birds has been seen feeding at the refuse tip. Cattle egrets will also gather in evening roosts with the larger egrets in mangroves and bushes around ponds such as at Caul's Pond, Little Harbour Ponds and Road Salt Pond although roost sites are prone to change.

LOCAL NOTES Cattle egret is a relative newcomer to Anguilla first arriving in the early 1960s.

YOUR NOTES

Green heron
Butorides virescens

HOW TO IDENTIFY IT   L 45 cm (18″). A small dark heron with a dark crown, white line below the eye, a long bill and striking yellow legs that are often seen trailing as it flies away when disturbed. Adults have a plain, rufous neck and this area and the breast is streaked brown on immature birds. It can be found around all the wetlands and occasionally on the coastal cliffs where they slowly and patiently wait for fish and other prey. A loud yelping *skyow* call often betrays its presence as it retreats into cover.

SIMILAR SPECIES   Adults are distinctive but immature birds are often seen at sites with young yellow-crowned night-herons. Night-herons are larger than green herons, a drabber grey-brown and have heavier, shorter bills.

WHEN TO SEE IT   You can find this resident, breeding bird in all months. The population is widespread but small and is likely to be close to 10 pairs.

WHERE TO SEE IT   Green herons can be found on all wetlands where there is fringing vegetation such as mangroves that provide a low perch for them to stalk prey. Regular sites include East End Pond, Caul's Pond, Bad Cox Pond, Little Harbour Ponds, Mead's Bay Pond and Road Salt Pond. They often feed in the open at the seaward end of Road Salt Pond.

LOCAL NOTES   At some coastal sites such as the eastern end of Mead's Bay green herons will creep Gollum-like down rocks to the point just above the wave limit and catch fish in the swell. They also occasionally perch on overhead wires by ponds.

YOUR NOTES

Yellow-crowned night-heron
Nyctanassa violacea

LOCAL NAME
Night gaulin

HOW TO IDENTIFY IT L 60 cm (24"). A medium-sized wading bird, adults are greyish with a bold black-and-white head pattern, yellowish legs and a dark, heavy bill. Immature birds are grey-brown with faint streaks on their neck and small pale spots on their closed wings. They are largely nocturnal birds and difficult are to see but occasionally can be found out in the open in the daytime around ponds or in sand dunes.

SIMILAR SPECIES See under green heron for identification of immature birds. Adults and young can be confused with the similar black-crowned night-heron although there are no recent records of this species.

WHEN TO SEE IT This is a resident breeding bird with a small population of fewer than 10 pairs.

WHERE TO SEE IT Roosting and feeding birds can be found in mangroves at sites such as Road Salt Pond, Blackgarden Bay Pond, Blowing Point Pond, Little Harbour Ponds and some of the smaller coastal ponds. Birds often feed on the coast and fly in to hunt in late afternoon off coastal rocks such as at Little Bay and the eastern end of Mead's Bay.

LOCAL NOTES These birds regularly fly into gardens after dark to feed on large insects and other prey. Listen out for a short, harsh *kowk*.

YOUR NOTES

White-cheeked pintail
Anas bahamensis

HOW TO IDENTIFY IT L 43 cm (17"). A bright, mottled brown duck, adults have a striking white cheek, a long white-edged tail and a bright red patch on the side of the bill. Pintails feed from the surface of ponds or by 'upending' to reach down into deeper water for plants when only the rear body and tail is visible. Small groups are often seen in low flight when a green patch or speculum is visible on the inner wing. This is the only duck likely to be seen on the island between May and August.

SIMILAR SPECIES Ruddy ducks have a white cheek but are much smaller, have a blue or grey bill and dive to feed. Females of other dabbling ducks lack the clear white cheek and are scarce visitors.

WHEN TO SEE IT This duck is present year round and will breed on many of the wetlands particularly when water levels are high. Ducklings have been seen in every month between November and August.

WHERE TO SEE IT It can be seen on most wetlands with the largest flocks typically found on East End Pond, Road Salt Pond and Caul's Pond.

LOCAL NOTES Its range is limited to the northern Caribbean and so is a local speciality for many visitors. Flocks of over 1,000 birds have been seen in the past but today the population is fewer than 200 pairs.

YOUR NOTES

Blue-winged teal
Anas discors

HOW TO IDENTIFY IT L 39 cm (15.5"). A medium-sized dabbling duck and surface feeder, males have a blue head with a large white crescent in front of the eye and a white patch on the rear body in front of a black tail. Females and immature birds are more difficult to identify and are mottled brown, have a faint dark line through the eye and a pale spot at the base of the bill. Both sexes have a pale blue panel at the front on the inner part of the upper wing easily seen in flight. They feed at the surface of the water and are often in the company of white-cheeked pintails.

SIMILAR SPECIES The lack of a pale white cheek and plain grey bill separate it from the resident white-cheeked pintails. Several other North American dabbling ducks have occurred, often among the flocks of blue-winged teals. Males of American wigeon, green-winged teal and others are easier to pick out but females require careful identification using a field guide.

WHEN TO SEE IT Teals are visitors from North America seen exclusively in the period September–April. In good years flocks of 100 to 200 birds can be found on some ponds.

WHERE TO SEE IT This species favours a few ponds for roosting with sheltered bays and cover such as mangroves. Bad Cox Pond, East End Pond, Little Harbour Ponds and Mimi Bay Pond are most often used although a few occasionally visit other ponds especially when they first arrive.

LOCAL NOTES This is the only dabbling duck species to regularly spend the northern winter on the island.

YOUR NOTES

Ruddy duck
Oxyura jamaicensis

HOW TO IDENTIFY IT L 38 cm (15"). This is the smallest of the visiting diving ducks and is compact with a large head and an obvious long tail. Males are chestnut coloured and have a striking white cheek on a dark head and a bright blue bill. Females and immature birds are duller, have a dark line across the pale cheek and grey bills. Small flocks regularly sit out in the open on Caul's Pond and intersperse bouts of diving with roosting when their heads are turned to rest on their backs and the long spiky tail is obvious

SIMILAR SPECIES The much larger white-cheeked pintail is the only other duck with a white cheek and these have a different shape and a red patch on the bill. Diving ducks such as ring-necked duck and lesser scaup visit the island occasionally and should be checked for carefully with a field guide.

WHEN TO SEE IT Most occur as visitors from North America in the October–April period although a family party with juveniles recently seen in August suggest these could be from a Caribbean breeding population. Flocks usually number fewer than 20 birds with 35 the peak count.

WHERE TO SEE IT The largely freshwater ponds of Caul's Pond and East End Pond are the only regular places to see this species.

LOCAL NOTES This species was first reported on Anguilla in 2000 and since then it has become a regular visitor.

YOUR NOTES

Osprey
Pandion haliaetus

LOCAL NAME
Fish hawk

HOW TO IDENTIFY IT L 58 cm (23"), WS 158 cm (63"). These are large and long-winged birds of prey, dark above and mostly white below with a dark eyestripe on a white head. They hold their wings slightly bowed in flight like a large gull. Look for their distinctive feeding action of hovering over water before dropping feet first to catch fish.

SIMILAR SPECIES The only large broad-winged bird of prey to be seen on the island.

WHEN TO SEE IT It is a visitor from North America with very small numbers passing through or staying on the island between October and March. They are most often seen singly but two birds have occasionally been seen together.

WHERE TO SEE IT The most reliable site to see ospreys is Caul's Pond where birds will fish and sit out in a belt of dead mangroves on the western shore. Long Pond, Road Salt Pond, and Little Harbour Ponds also attract fishing birds but otherwise they can occasionally be seen flying anywhere across the island or fishing along the coast.

LOCAL NOTES Ospreys will take prey to a perch to feed. Where bushes are absent, they will use the tops of utility poles.

YOUR NOTES

Common moorhen
Gallinula chloropus

HOW TO IDENTIFY IT L 35 cm (14"). A small water bird, moorhens are generally dark brown above and grey below with a white undertail, a white stripe along its flanks and a colourful yellow-tipped red bill. Immature birds are duller and lack the red on the bill. They feed on the surface of ponds or walks around banks when their yellow-green feet and long toes can be seen. Moorhens can be very vocal and have a short clucking call and a distinctive long ringing set of calls that slows at the end.

SIMILAR SPECIES This species often occurs with coots and these are dark grey-black, have smaller white patches under their tail and strong and striking white bills with a white shield on the forehead. The scarcer and more furtive sora rail is brown above with white flecks and has brown and white barring on the lower flanks. Adult sora rails are grey on the face and upper breast, have a yellow bill and a dark patch in front of the eye.

WHEN TO SEE IT Birds are present year round and this is a common breeding bird on some ponds especially when water levels are high. The peak breeding population is likely to be up to 50 pairs.

WHERE TO SEE IT It is easiest to see on the more freshwater wetlands at East End Pond, Bad Cox Pond and Caul's Pond where large numbers occur following a successful breeding season. A few can be seen on other ponds with grassy shorelines such as Road Salt Pond.

LOCAL NOTES Look out for families of small fluffy black chicks during the breeding season.

YOUR NOTES

American coot
Fulica americana

HOW TO IDENTIFY IT L 38 cm (15.5"). This small water bird looks all dark at distance except for a striking white bill and white shield on the forehead. Coots are usually seen on the surface of ponds close to emergent vegetation where they dive for aquatic plants surfacing close to the point of the dive. At close quarters thin white lines are visible on the under tail.

SIMILAR SPECIES See text for common moorhen. Caribbean coot is very similar but typically has a swollen white frontal shield that stretches well onto the forehead above the eye. This species is very difficult to identify with certainty as it will hybridise with American coot. Caribbean coot has only been seen on East End Pond in recent years and remains a threatened species throughout its range.

WHEN TO SEE IT Recent sightings have shown it to be present year round although scarce in the breeding season from April to August. This species and Caribbean coot have bred in recent years on East End Pond, the chicks having paler heads than moorhen chicks. More birds arrive from North America in the period October–March when small flocks will gather.

WHERE TO SEE IT Coots are only found with any regularity on East End Pond and Caul's Pond and these have areas of emergent vegetation and cover in shallow bays.

LOCAL NOTES In dry periods when favoured freshwater sites dry out, coots appear to move off the island completely.

YOUR NOTES

Black-bellied plover
Pluvialis squatarola

HOW TO IDENTIFY IT L 28 cm (11.5"). This is a large shore bird, speckled grey above and paler below with a large eye, short dark bill and dark legs. In flight look for a white wing stripe and tail and a striking black patch where the underwing meets the body. The call is a clear, slightly mournful three-note whistle. Plovers feed by sight, standing motionless looking for invertebrate prey before making a short run to catch it.

SIMILAR SPECIES The combination of large size and short bill separate it from all species except the American golden plover. This scarce visitor in September–November is slightly smaller and has a darker crown with a white brow behind the eye. It is also more neatly spotted above, is barred and spotted on the breast and looks leggier. It also has plain grey underwings without the black patch.

WHEN TO SEE IT A few pass through on migration from the Arctic from August onwards with small numbers then present throughout September to May. Fewer than 10 birds is a typical count but roosts have reached over 60 birds on some ponds.

WHERE TO SEE IT They are seen most often at Long Pond and Blowing Point Pond but singles can turn up at most wetlands including the edges of ponds that are drying out.

LOCAL NOTES This species is typically seen in non-breeding plumage but a few arriving in August or staying into May develop their superb black-bellied breeding plumage.

YOUR NOTES

Snowy plover
Charadrius alexandrinus

HOW TO IDENTIFY IT L 15 cm (6"). This is a striking little shore bird with an endearing expression. A small plover, it has very pale grey upperparts, white underparts and grey patches on the crown, face and the side of the breast. Males have black head and breast markings during the breeding season. It will run rapidly across sand and mud on pale grey legs.

SIMILAR SPECIES See under semi-palmated plover for comparison with small plovers. The very rare piping plover has been seen with flocks of snowy plovers and this is also very pale with a tiny black bill and orange not grey legs. Sanderlings also run rapidly across the sand but are more often on open beaches and are larger, have black legs, a black patch on the front of the closed wing, no breast patch and a longer bill.

WHEN TO SEE IT A few pairs, probably less than 10, breed and small numbers are present year round.

WHERE TO SEE IT Snowy plovers are scarce on the islands but can be found on sand spits, beaches and the muddy edges of coastal ponds. Long Pond and West End Pond are the most regular sites.

LOCAL NOTES This species and other plovers will lay their eggs in shallow scrapes within dry sandy areas. Try to avoid disturbance to potential nesting sites during April–July and inadvertent trampling of the wonderfully camouflaged eggs and nest.

YOUR NOTES

Wilson's plover
Charadrius wilsonia

HOW TO IDENTIFY IT L 19 cm (7.5"). This plover is pale brown above and white below with a single brown breast band, pale brown face patch and crown. Its large, heavy black bill is distinctive. The male has darker head markings and a black breast band when in breeding plumage. Its feeding behaviour is similar to other small plovers. It can be easily overlooked as the pale upper parts often match its surroundings and they often sit or stand motionless.

SIMILAR SPECIES See under semi-palmated plover.

WHEN TO SEE IT It can be seen year round and breeds from April to July in small numbers, probably fewer than 10 pairs, on sand bars and sandy edges of ponds.

WHERE TO SEE IT In the breeding season it is more likely seen on the big coastal lagoons such as Long Pond, Cove Bay Pond, Mead's Bay Pond and West End Pond. Outside the breeding season a few birds can also be seen at other ponds especially where water levels are dropping. A few can also be found on quieter rocky and sandy beaches.

LOCAL NOTES Mixed flocks of plovers will gather especially to roost when parties of up to 24 Wilson's plovers have been reported.

YOUR NOTES

Semi-palmated plover
Charadrius semipalmatus

HOW TO IDENTIFY IT   L 18 cm (7"). A small shore bird, brown above and white below it has a single bold, complete breast band. The head is well marked with a dark cheek patch and crown and a white forehead. The short bill is dark in non-breeding birds and orange-based in adults. It has pale, orange legs and like all the small plovers, it has a white wing bar and a dark tail. This species is typically found in open muddy areas around bays and wetlands.

SIMILAR SPECIES   One of four small plovers of which Wilson's plover is closest in markings and this is best separated by its large and heavier black bill, paler brown upperparts and pale, dull leg colour. Killdeers are distinctive in having two breast bands and longer paler legs. Snowy plovers are the smallest and look very pale with just a black or grey breast patch, lack a full breast band and have grey legs.

WHEN TO SEE IT   It arrives on migration from Arctic Canada from August onwards and is then present in small numbers through to May. It often gathers in small flocks if feeding conditions are suitable when peak counts can reach 80.

WHERE TO SEE IT   Coastal lagoons with muddy shores attract most birds such as West End Pond, Long Pond, Road Salt Pond, Cove Bay Pond and Blowing Point Pond. A few occur on other ponds and occasionally on the quieter sandy beaches such as at Crocus Bay.

LOCAL NOTES   This bird is often found in small discrete flocks or in groups within larger mixed flocks of shore birds. It can be picked out from other shore birds at a distance on muddy edges of pools by its small size, upright stance and feeding action with its short runs for prey.

YOUR NOTES

Killdeer
Charadrius vociferus

HOW TO IDENTIFY IT L 25 cm (9.5"). This is the largest of the banded plovers and looks longer-legged than the rest. The double dark breast bands are distinctive. The head pattern and white bar on the upperwing are similar to other plovers but the black-and-white tail contrasts with a warm orange rump. Killdeers draw attention to themselves as they are more vocal than other plovers with high-pitched notes including a phrase that led to its common name.

SIMILAR SPECIES See under semi-palmated plover.

WHEN TO SEE IT A few are present year round with a small, local breeding population joined by migrants from North America in the September–April period. Look out for chicks and juvenile birds in April–June. They will gather in small flocks usually involving less than 10 birds.

WHERE TO SEE IT Killdeers are more widespread than other plovers feeding on grassy areas including the airport grasslands and cricket pitches such as at Road Salt Pond and the landward end of Long Pond. They breed on bare areas around ponds with East End Pond and Road Salt Pond particularly favoured. In the period September–March, a few can turn up on beaches and at most ponds often away from other shore birds.

LOCAL NOTES After rain, they often feed on small wet flashes on grassland or by puddles in bare areas such as the western end of Sandy Ground.

YOUR NOTES

Black-necked stilt
Himantopus mexicanus

LOCAL NAME
Pond bird

HOW TO IDENTIFY IT L 38 cm (15"). An unmistakable bird, stilts have a combination of black upperparts, white underparts, a long fine bill and very long pink-red legs. Stilts feed around the edges of ponds or wade in deeper water to pick insects from surface or catch fish and small crabs. They are very vocal and the typical call is a noisy *wit, wit, wit*.

WHEN TO SEE IT A resident species it also breeds on most ponds in varying numbers with 150–200 pairs present in good years. The breeding population appears to fluctuate in response to water levels in ponds and it can be absent from favoured ponds if they dry out.

WHERE TO SEE IT A few can be found on most ponds at any time but the largest numbers are found on West End Pond, Road Salt Pond and East End Pond and the birds here can be wonderfully approachable at times.

LOCAL NOTES This species is hard to miss on Anguilla and is the only shore bird breeding in large numbers. Stilts can be very noisy giving incessant loud, sharp calls particularly from protective adults and they will fly directly at you if you get too close to the chicks.

YOUR NOTES

Greater yellowlegs
Tringa melanoleuca

HOW TO IDENTIFY IT   L 35 cm (14"). This is a large, slender shore bird with grey speckled upperparts, a lightly streaked neck and breast and white underparts. It has distinctive long yellow legs and a long, greyish, slightly upturned bill. The typical call is a ringing *tew tew tew*.

SIMILAR SPECIES   Shore bird identification can be very tricky and it is not easy to separate this species from lesser yellowlegs. Greater yellowlegs is obviously larger when the two are together, its bill, longer, paler at the base and slight upturned, is a good feature. The bill of lesser yellowlegs is comparable in length to the width of its head but with greater yellowlegs the bill is obviously longer than the head. Greater yellowlegs also looks longer in the leg between the 'knee' joint and the body. Lesser yellowlegs typically has a softer one or two note call.

WHEN TO SEE IT   A non-breeding visitor from North America, a few can be seen year round. The largest numbers occur when passage birds join the wintering population in the September–April period. Fewer than 10 birds are usually present at a site although exceptionally up to 29 have been seen together.

WHERE TO SEE IT   A few will be present wherever there are large flocks of shore birds feeding together such as East End Pond, Long Pond, Road Salt Pond and West End Pond. Most ponds will attract this species at times either singly or in small groups when its call is often the first indication of its presence.

LOCAL NOTES   Anguilla is a great place to watch shore birds and as both species of yellowlegs are often together, it is worth spending time looking at the subtle differences.

YOUR NOTES

Lesser yellowlegs
Tringa flavipes

Pond dipper

HOW TO IDENTIFY IT L 26 cm (10.5"). A medium-sized, slender wading bird it has long yellow legs and a fine, dark grey bill. The upperparts are grey and lightly speckled and the breast has fine grey markings on otherwise white underparts. The face is neatly marked with a dark line through the eye and a thin white 'eyebrow' in front of the eye. Look for grey upperparts in flight contrasting with a square white rump and lightly barred tip to the tail. They will feed out in the centre of ponds where they swim readily. This is a vocal species and will often give a single *chew* call.

SIMILAR SPECIES See text under greater yellowlegs.

WHEN TO SEE IT A few are present year round but large flocks can occur on migration and in the period August–May when it is the commonest shore bird on the island. When water levels are high up to 250 birds will congregate on some ponds. If an abundance of food coincides with peak migration even larger flocks occur with a peak count of over 600 birds.

WHERE TO SEE IT They are present on most of the ponds and wetlands at some time, usually in small groups and will readily congregate in larger flocks if water levels are suitable on Road Salt Pond, Long Pond, West End Pond and East End Pond.

LOCAL NOTES Large flocks of greyish wading birds feeding out in the centre of ponds are likely to be this species or stilt sandpipers.

YOUR NOTES

Willet
Catoptrophorus semipalmatus

HOW TO IDENTIFY IT L 39 cm (15.5"). A large wading bird, it is heavier than the yellowlegs with plainer grey upperparts and a grey wash across the sides of the breast. Look out for the long grey legs and thick grey bill that is paler at the base. The relatively obscure plumage changes in flight when they show striking black wings with a broad white bar across the upperwing and underwing. In April and May listen out for its song delivered in a display flight, a clear *will will willet.*

WHEN TO SEE IT A few birds can be found in most months but is much easier to see in the April–August period when up to 20 pairs breed at coastal sites.

WHERE TO SEE IT East End Pond and Long Pond attract feeding birds during the April–June breeding season. At other times, these ponds, Little Harbour and Road Salt Pond are regular sites but individuals could turn up on any of the ponds.

LOCAL NOTES Willets nest in quiet, sandy areas and care should be taken to avoid disturbance in the breeding season.

YOUR NOTES

Spotted sandpiper
Actitis macularia

HOW TO IDENTIFY IT L 18 cm (7.5"). A small short-necked, short-legged wading bird usually found singly around pond edges. It is easy to identify when in its spotted plumage as most adult birds are in April–May and July–August. At other times it is plain brown above, white below with brown breast patches and a thin white eye-stripe. It also has a pale short bill, pale legs and a short white stripe on the upperwing. Look for the characteristic bobbing motion when feeding and bursts of shallow stiff wing beats followed by glides as it flies low over the water. It often calls in flight with two to three high-pitched ringing whistles *we we weet*.

SIMILAR SPECIES Solitary sandpiper is darker above with a plain upperwing and underwing, a dark-centred white tail and a pale ring around the eye. A few solitary sandpipers pass through the island in August–October but are more often found in temporary pools and grassy areas.

WHEN TO SEE IT A migrant from North America, it can be seen in small numbers from July–May with typically no more than five to six birds spread around the edges of favoured ponds.

WHERE TO SEE IT Look for them on any pond with a muddy or rocky edge but favoured ponds include Road Salt Pond, Rendezvous Bay Pond and West End Pond.

LOCAL NOTES When walking or driving around the larger salt ponds you regularly come across single birds sitting on a rock or standing still by the water's edge. Once they relax they resume the usual bobbing action when walking.

YOUR NOTES

Whimbrel
Numenius phaeopus

HOW TO IDENTIFY IT L 42 cm, (17"). This is one of the largest wading birds and looks brown at a distance with a pale belly, grey legs and a distinctive long down-curved bill. With a closer view its striped head is visible with dark stripes through the eye and high on the side of the crown. Whimbrels often call when disturbed giving a loud series of whistles. They will occasionally perch on low branches by pond edges.

SIMILAR SPECIES No other curlew species have been seen to date.

WHEN TO SEE IT The first birds arrive from breeding grounds high in North America from July and a few are then present most months through to May. It is mostly seen singly but small groups of up to five to six are also seen.

WHERE TO SEE IT It is most often found on Caul's Pond, East End Pond, Blowing Point Pond and Little Harbour Ponds where there is surrounding cover, exposed mud and shallow water for feeding.

LOCAL NOTES The American race found on Anguilla has a brown rump unlike populations in Europe and Siberia with a white rump.

YOUR NOTES

Ruddy turnstone
Arenaria interpres

HOW TO IDENTIFY IT L 21 cm (8.5"). A small short-legged, short-necked wading bird, it looks very dark above and white below at distance. The head pattern is bold especially in breeding plumage and links to dark patches on the breast. Its legs are orange and the short bill is dark grey. Turnstones have a striking and distinctive pattern in flight with white stripes along the upperwing and where the wing meets the body, and a black band near the tip of the white tail. The flight call is a low *tuk a tuk*. Turnstones will turn over stones in a search for shrimps and other small creatures.

WHEN TO SEE IT It is most frequent in the period August–May as birds arrive or pass through on migration from breeding grounds high in North America and is typically found in groups of fewer than 20 birds.

WHERE TO SEE IT Ponds with rocky and muddy edges are favoured especially Road Salt Pond, Long Pond and West End Pond although small numbers will turn up at most ponds. They are also typically found with sanderlings on sandy beaches and rocky or remote coasts such as at Forest Bay, Island Harbour and Savannah Bay.

LOCAL NOTES Turnstones will occasionally rest on small moored boats off Sandy Ground or Shoal Bay East. They also feed or rest on the low salt pond walls found at Road Salt Pond and West End Pond. In May and August you may see adults in their lovely chestnut and black pattern that gives it their common name

YOUR NOTES

Sanderling
Calidris alba

HOW TO IDENTIFY IT L 20 cm (8"). Sanderlings are small pale-grey wading birds with black legs and a short black bill. Often a black crescent is visible on the front of the closed wing. Young birds are darker above with fine dark lines in a patch on the side of the upper breast. They will feed with other birds especially ruddy turnstones but they also have a highly individual feeding action on sandy beaches running very fast between waves to pick invertebrate prey from the surf.

SIMILAR SPECIES Semi-palmated sandpipers in non-breeding plumage are smaller, more compact and slightly darker above. A single sanderling with other wading birds on the ponds can prove surprisingly tricky to identify in some plumages!

WHEN TO SEE IT Small numbers arrive from July with birds present then through to May. Flock size is mostly less than five with parties of up to 25 birds the exception.

WHERE TO SEE IT Look for this bird feeding among flocks of ruddy turnstones on beaches and rocky shores, Island Harbour and Savannah Bay are particularly good for this species. Birds will also join feeding flocks on the coastal lagoons especially West End Pond and Long Pond.

LOCAL NOTES A classic location for wave running behaviour is late afternoon at Mead's Bay where one or two birds often entertain.

YOUR NOTES

Semi-palmated sandpiper
Calidris pusilla

HOW TO IDENTIFY IT L 15 cm (6.25"). A tiny wading bird, it is grey above and white below save for a grey breast patch. Its short fine bill and legs are black. A low *chirrp* call can be heard in flight when you can also see a thin white bar across the upperwing. They are usually seen in flocks and most often feed in shallow water or by muddy edges of ponds.

SIMILAR SPECIES Least sandpipers are similar in size but generally darker browner above, with a darker crown and a slightly down-curved and fine-tipped bill. Least sandpipers also have paler, yellowish legs, often appearing more angled towards the front of the body from the visible joint. Least sandpipers more often feed in drier areas away from the ponds although this is not a consistent feature. More tricky to separate from semi-palmated sandpipers are the few western sandpipers that pass through on migration. Use a field guide to check these and see text for the slightly larger white-rumped sandpiper.

WHEN TO SEE IT A few are likely to be seen in all months but the main arrival from North America begins in July with a return migration in April–May. Flocks of up to 100 birds can occur in all these months especially in shallow bays with receding water levels. When the feeding conditions are ideal up to 270 have been seen together.

WHERE TO SEE IT This is a widespread visitor to the islands and is one of the commonest wading birds turning up on any wetland with shallow water or muddy edges. The largest flocks occur on Long Pond, West End Pond, Road Salt Pond, Blowing Point Pond and East End Pond.

LOCAL NOTES Look out for birds in juvenile plumage arriving with adults in August.

YOUR NOTES

Least sandpiper
Calidris minutilla

HOW TO IDENTIFY IT   L 15 cm (6"). A tiny wading bird generally brown above with a dark brown streaked crown and white below with a short thin dark bill with a slight but discernable down-curve. The greenish yellow legs are distinctive and it has a creeping feeding action with angled legs.

SIMILAR SPECIES   See under semi-palmated sandpiper.

WHEN TO SEE IT   This species migrates south from North America with the first birds arriving from late July onwards and leaving by May. Small flocks will stay throughout this period. They generally occur in smaller flock sizes than semi-palmated sandpiper with a recent peak of 93 birds.

WHERE TO SEE IT   Least sandpipers can be found around muddy edges of any pond or seasonal wetland and are occasionally seen in drier areas. The largest flocks occur on Long Pond, West End Pond, Road Salt Pond, Blowing Point Pond and East End Pond.

LOCAL NOTES   These tiny waders or peeps can be very difficult to identify but this species and semi-palmated sandpiper will often feed in the same area providing excellent and often close opportunities to separate the two.

YOUR NOTES

White-rumped sandpiper
Calidris fuscicollis

HOW TO IDENTIFY IT L 19 cm (7.5"). This is a small wading bird with grey upperparts including the head and neck, a neat white stripe above the eye, small grey bill and grey legs. Strong streaking on the upper breast gives an impression of a dark 'front' to the bird. Juvenile birds are brighter with pale edges to the feathers of the back and wings. It can be picked out in mixed flocks of shore birds by its different shape with heavier head and neck and long wings given a tapered look to rear. In flight the white rump is diagnostic and they also have a white wing bar.

SIMILAR SPECIES This species is slightly but obviously larger than tiny wading birds such as semi-palmated sandpiper. Sanderlings are slightly larger and paler. The pectoral sandpiper occurs from August–November and is slightly larger, browner with a densely streaked breast that ends abruptly in a neat line. These lack the white rump, have a subdued wing bar and are typically found in more grassy pond margins or pools on grassland. Other similar small wading birds could occur but these are rare on Anguilla and should be checked in a field guide.

WHEN TO SEE IT All recent sightings are in the August–December period suggesting they fly further south on continuing migration from North America. They are most often seen singly but small flocks of up to seven birds are often seen and a peak count of 25 birds was exceptional.

WHERE TO SEE IT This species could occur on most wetlands with recent records from West End Pond, East End Pond, Road Salt Pond, Long Pond and Mead's Bay Pond.

LOCAL NOTES Most records are of single birds or small groups picked out as 'different' in size and shape when checking through flocks of feeding shore birds.

YOUR NOTES

Stilt sandpiper
Calidris himantopus

HOW TO IDENTIFY IT   L 21 cm (8.5"). A small wading bird that looks bigger because of its long, yellowish-green legs. In non-breeding plumage it is greyish with a dark line between the eye and bill and a pale stripe above the eye. They have a proportionately long and down-curved black bill. The feeding action is distinctive with flocks feeding together in open water making a jabbing action for worms and other prey often submerging their head, neck and shoulders. Adult birds in breeding plumage with strong dark barring on their underparts can be seen during April–May and July–August.

SIMILAR SPECIES   On Anguilla, the stilt sandpiper is often present with both yellowlegs species and is smaller than these with its distinctive slightly down-curved bill. Its feeding action also differs and is more jabbing in deep water.

WHEN TO SEE IT   It arrives from breeding grounds high in the Arctic and varying numbers can be seen in all months with most in the period August–May. After heavy rains large flocks will congregate on some ponds involving 300–450 birds.

WHERE TO SEE IT   This species feeds mostly in deeper water so look for ponds with higher water levels. Favoured ponds are Long Pond, East End Pond, Road Salt Pond and West End Pond but it could turn up on most ponds.

LOCAL NOTES   If conditions are right, stilt sandpipers form large flocks and gather away from the water's edge to feed with flocks of lesser yellowlegs. When ponds are full, they will also feed by the pond edges providing close views.

YOUR NOTES

Short-billed dowitcher
Limnodromus griseus

HOW TO IDENTIFY IT L 27 cm (11"). A medium-sized wading bird with short legs and a proportionately very long bill. Adults are greyish above and pale below with streaking on the breast sides and barred flanks and undertail. Dowitchers also have a dark line through the eye with a pale stripe above. In flight, they show a white oval patch on the back and a white trailing edge to the inner wing. They tend to form small discrete flocks within larger flocks of shore birds and feed by probing vigorously into mud and grassland in a search for prey items.

SIMILAR SPECIES Wilson's snipe is the only medium-sized wading bird with a comparable long bill. This is browner with heavily barred flanks and pale lines on the back. Juvenile dowitchers (August–November) are browner but lack the bold back stripes and boldly marked head of Wilson's snipe. Long-billed dowitcher is very similar but is a rare visitor to Anguilla. It typically has a longer bill and has bolder dark bars than white bars on its tail. A field guide is recommended to separate the two species.

WHEN TO SEE IT This species arrives in small numbers from North America in August with a few present each month through to May. Typically they are found in flocks of fewer than 10 with a recent peak count of 16 birds.

WHERE TO SEE IT Dowitchers feed in shallow water and muddy and grassy edges at most of the larger ponds especially, Road Salt Pond, Long Pond, West End Pond and East End Pond.

LOCAL NOTES Look for juvenile birds arriving with adults in August.

YOUR NOTES

Belted kingfisher
Ceryle halcyon

HOW TO IDENTIFY IT L 30 cm (12″). This is a striking, large blue and white bird with a strong dark bill. The head and upperparts are blue with a white collar and a shaggy crest obvious at times. The underparts are white except for a blue breast band with females having an additional chestnut band below this. They are very vocal with the typical call being a loud chattering rattle. Kingfishers look for prey from a conspicuous perch or by hovering before diving into ponds and shallow seas.

WHEN TO SEE IT This is a non-breeding visitor in small numbers and is only seen from September to April.

WHERE TO SEE IT It is mostly a coastal bird and can be found feeding among mangrove-fringed coastal lagoons and in the surf of sheltered bays. The best sites are Little Bay, Little Harbour Ponds, Blowing Point Pond, Mead's Bay, Cove Bay Pond and Rendezvous Bay Pond.

LOCAL NOTES To date all records separated by sex have been males suggesting these travel further from the breeding grounds than females.

YOUR NOTES

Other water birds

Little blue heron
Egretta caerulea
A rare visitor that will often stay several months once it reaches the islands. The only recent record is of an adult in 2000.

Tri-coloured heron
Egretta tricolor
A scarce visitor with one to two birds in the period October–January at least with both adults and first-year birds seen in recent years. This North American species appears to be increasing on Anguilla.

Glossy ibis
Plegadis falcinellus
A rare visitor, a single bird was seen at Road Salt Pond in 2002.

Greater flamingo
Phoenicopterus ruber
A rare visitor with a single immature bird present for several months at Cove Bay Pond in 2000–2001, the first record since two birds stayed for three years in the 1980s.

West Indian whistling duck
Dendrocygna arborea
A rare visitor from populations within the Caribbean, a party of five birds was seen briefly at East End Pond in December 2003.

Snow goose
Chen caerulescens
A rare visitor with a single record of a dark phase adult seen at East End Pond over four months from 2004.

Green-winged teal
Anas crecca
A rare visitor from North America it is seen most years with flocks of up to seven birds occurring in the period October–March. They favour ponds with cover at the edge and are most often found when checking through flocks of blue-winged teals.

Northern pintail
Anas acuta
A rare visitor, there has been a single record of two birds seen at Grey Pond and Caul's Pond in January–February 2000.

Northern shoveler
Anas clypeata
A rare visitor, two single birds have been seen, an immature male at Caul's Pond in January 2001 and an adult male at East End Pond and Mimi Bay Pond in February 2006.

American wigeon
Anas americana
A rare visitor this species is seen in most years in the period December to February. It typically arrives in parties of fewer than five birds although a flock of 18 birds was seen in 2000.

Ring-necked duck
Aythya collaris
A rare visitor seen on the more freshwater wetlands of Caul's Pond and East End Pond in the period December–April. This diving duck occurs in small flocks of up to 10 birds.

Lesser scaup
Aythya affinis
A rare visitor, this North American diving duck is seen in most years in the November–April period with Caul's Pond the most favoured site. It typically arrives in small flocks although 22 birds were seen in 2000.

Hooded merganser
Lophodytes cucullatus
A rare visitor this diving duck has only been seen on one occasion with two females at Road Salt Pond from January to February 2000.

Sora rail
Porzana carolina
A scarce visitor to mangroves and overgrown margins to ponds, up to five a year of these small rails are seen in the October–February period. Both adult and first-year birds have been seen annually since the first reports in 2000 although their secretive nature suggests they have been overlooked in the past. Favoured ponds include Blackgarden Bay Pond, Cove Pond, East End Pond and Little Harbour Ponds.

Caribbean coot
Fulica caribaea
A scarce visitor and rare breeding bird with one to two birds seen at East End Pond annually since 2000 with more records in the breeding season from March–July.

This rare and declining species is largely restricted to the Caribbean where it is declining. American coots have bred on Anguilla recently and the two coot species will inter breed and hybridise, making it increasingly difficult to identify Caribbean coots with certainty.

American golden plover
Pluvilais dominica
A scarce visitor, this North American shore bird stops off in small numbers in the October–November period with the irrigated grasslands of the golf course likely to be the best place to see this species.

Piping plover
Charadrius melodus
A rare visitor with two individuals joining the plover flocks in the period October–January. The first was seen on the beach at Long Pond in December 2000 and returned for two further winter periods. The second bird was in the same area in October 2006.

American oystercatcher
Haematopus palliatus
A resident bird it breeds in small numbers on the undisturbed coastline of offshore cays including Scrub Island with fewer than 10 pairs present. Outside the breeding season oyerstercatchers will frequent rocky coasts such as off Long Pond, and the stretch from Katouche Bay to Limestone Bay where they can be surprisingly elusive.

American avocet
Recurvirostra americana
A rare visitor, two birds were seen on Sandy Island and Scrub Island in June 1999.

Solitary sandpiper
Tringa solitaria
A scarce visitor, small numbers pass through from August–November on southerly migration from North America. Typically seen singly this species is found on temporary pools on grassland or among vegetation on muddy margins of ponds such as at Bad Cox Pond.

Hudsonian godwit
Limosa haemastica
A rare visitor there is one record of a single juvenile bird present at Road Salt Pond and other locations in October 2006.

Red knot
Calidris canutus
A scarce visitor, this species could occur in any

month from August–June. Recent records have been from Long Pond, West End Pond and Windward Point.

Western sandpiper
Calidris mauri
A rare visitor in recent years that is probably over-looked as it is very similar to semi-palmated sandpiper. One or two birds probably occur annually among flocks of small shore birds with Long Pond and West End Pond the most favoured ponds.

Pectoral sandpiper
Calidris melanotus
A scarce visitor, it is seen singly or in small flocks on southerly migration in the period August–November. This species favours grassy margins to ponds such as Road Salt Pond with larger flocks of up to 20 birds occasionally gathering on temporary floods on grassland such as at Wallblake airport or on the fairways of the golf course.

Curlew sandpiper
Calidris ferruginea
A rare visitor from Eurasia with a single bird reported in breeding plumage at Long Pond in June 2000.

Ruff
Philomachus pugnax
A rare visitor from Eurasia, a single bird was found among the flocks of yellowlegs at Mead's Bay Pond in April 2000.

Long-billed dowitcher
Limnodromus scolpaceus
A rare visitor this species is difficult to pick out from short-billed dowitcher and there is a single sighting of three birds at West End Pond in September 1994.

Wilson's snipe
Gallinago delicata
A scarce visitor to grassy areas and margins of ponds with small numbers involving one to four birds reported in the period September–January. This easily overlooked species was formerly known as common snipe.

Wilson's phalarope
Phalaropus tricolor
A scarce visitor, it may be present most years with September likely to be the best month to see it. Typically found singly there is a remarkable record of a flock of 31 birds at Cove Bay Pond in September 1993.

American kestrel
Falco sparverius

LOCAL NAME
Killy-killy

HOW TO IDENTIFY IT L 23 cm (9") WS 55 cm (22"). Kestrels are small, long-winged, long-tailed birds of prey. Two vertical dark lines on the head are a good field feature. When close by they appear brightly coloured with a dark barred chestnut back and darker marks on a pale breast. Males have neater dark spots on the breast and spotted grey upperwings. Its ringing *killi-killi* call gives rise to its local name. Kestrels fly with shallow, flicking wing beats and glide on stiff wings low over bushes. They are often encountered sitting on utility poles or wires from where they will drop on insects or make long straight glides after prey.

SIMILAR SPECIES The merlin is a little heavier, is darker, plainer brown or grey above, and has weaker face markings and stronger darker streaks on its breast. Merlins are a scarce visitor in small numbers and more often seen in open areas around ponds than on wires among housing and buildings.

WHEN TO SEE IT This is the only bird of prey present year round and breeds across the island in small numbers with a population of less than 25 pairs.

WHERE TO SEE IT It is easy to see across most of the mainland in open areas and grassland where it readily flies to a vantage point such as wires or the top of a utility pole.

LOCAL NOTES Hunting birds will often sit out feeding on a large insect or lizard allowing close views of the beautifully marked plumage.

YOUR NOTES

Zenaida dove
Zenaida aurita

LOCAL NAME
Turtle dove

HOW TO IDENTIFY IT L 25 cm (10"). This is the commonest medium-sized dove and is brown above with dark spots on the wing, and pink-brown below. Adults have a small dark crescent below the eye and a shiny purple patch on the side of the lower neck. A white trailing edge to the inner part of the upperwing and white edges to the tail are good features on flying birds. Immature birds are plainer coloured. The soft cooing calls are one of the background sounds of the island.

SIMILAR SPECIES White-winged dove is also seen on overhead wires and is paler brown with a striking white crescent along the front edge of the closed wing. A few feral rock doves or town pigeons occasionally occur and these typically have a dark grey head and neck and paler grey wings with two dark bands across them. Eurasian collared dove, an introduced species, has been seen on Anguilla and is established at nearby St Martin and Saba. These are paler and have a neat black collar on the hind neck. Check a field guide for further information on pigeons and doves.

WHEN TO SEE IT It is common in all months and a familiar resident and breeding bird. It forms small flocks of up to 25 birds outside the breeding season. Small numbers are also present on Little Scrub and Scrub Island.

WHERE TO SEE IT Zenaida doves are widespread around habitation and open areas, readily perch on overhead wires and are easily seen across the island when travelling around.

LOCAL NOTES They were once hunted on the island but are now a common and tame bird around settlements and throughout the scrub. The turtle dove was voted Anguilla's national bird in 1993.

YOUR NOTES

Common ground-dove
Columbina passerina

HOW TO IDENTIFY IT L 15 cm (6"). A tiny, pale grey dove with dark spots on the upperwing and scaly markings on the neck and upper breast. This species is more colourful in flight with a rufous patch on the outer wings and white corners to a black tail. Ground-doves usually fly low to the ground and have audible wing beats. They are often found on the ground in small flocks where they tend to be a little nervous continually flying ahead of the observer back to the ground or to a low perch in a bush.

SIMILAR SPECIES No other small doves occur.

WHEN TO SEE IT They are present year round breeding in scrub across the island and are a little easier to see in the main breeding period from April–September.

WHERE TO SEE IT This species can be scarce at times but is fairly easy to find on a walk through areas of low scrub particularly around Caul's Pond, Savannah Bay, Auntie Dol Bay and Long Pond.

LOCAL NOTES Ground-doves will build their open nests of grasses and twigs on or close to the ground as well as in bushes.

YOUR NOTES

Mangrove cuckoo
Coccyzus minor

Soldier bird

HOW TO IDENTIFY IT L 27 cm (11"). A medium-sized long-tailed land bird, it is a sleek grey-brown above and off-white below with pale orange on the lower belly. The crown is grey with a black triangular mask narrowing behind the eye. The black undertail is often fanned showing a row of bold white spots on each edge. The bill is short and curved with a yellow lower mandible. Cuckoos move slowly through mangroves, trees and scrub stopping to search for prey often twisting their neck to look upwards into branches. Their call is a low croak, *kwok ok ok ok* carrying over short distances.

SIMILAR SPECIES Migrant yellow-billed cuckoos pass through during September–November. These lack the black mask, are whiter below and have a rufous patch on the upperwing that can also be seen on the closed wing.

WHEN TO SEE IT A resident bird, it breeds in small numbers across the island.

WHERE TO SEE IT It can be seen around ponds and in scrub and gardens across the island. Look for them around Caul's Pond, Cove Bay Pond, Little Harbour Ponds, Junk's Hole Pond and West End Pond.

LOCAL NOTES Often the low croaking and repeated call note is the only sign it is present but with patience you can often be rewarded with good views as they move slowly through vegetation.

YOUR NOTES

Green-throated carib
Eulampis holosericeus

HOW TO IDENTIFY IT L 11 cm (4.5"). A small and fast-moving hummingbird, looking all metallic green at distance but with close views it has a metallic blue patch at the base of its green breast and a darker belly and tail. The long, thin down-curved black bill is distinctive. It actively feeds from flowers hovering to insert its bill and flying backwards before moving on.

SIMILAR SPECIES The only other hummingbird is the Antillean crested hummingbird which is smaller and has a short crest and shorter straight bill. Female and immature crested hummingbirds have a pale breast.

WHEN TO SEE IT A resident, breeding bird, it is seen in all months but is more obvious when flowers are in bloom. The population is difficult to estimate being spread sparsely throughout suitable habitat but is likely to be at least 25 pairs.

WHERE TO SEE IT It can be seen across the island in scrub and gardens. The areas around Caul's Pond, Island Harbour and West End Pond are usually productive but they can often be seen in gardens elsewhere.

LOCAL NOTES This is one of the highlights of any birdwatching visit. It is a Caribbean specialty found from Puerto Rico eastwards and down through the Lesser Antilles. On Anguilla it is often found feeding on flowering white cedar trees.

YOUR NOTES

Caribbean elaenia
Elaenia martinica

LOCAL NAME
LOCAL NAME
Weave

HOW TO IDENTIFY IT L 16 cm (6.5"). This is a small pale bird, grey-brown above and pale below with a whitish throat and belly and a short thin dark bill. Two whitish wing bars are obvious and it occasionally displays a small crest on its crown with a yellowish streak in the middle. Very vocal, it throws its head back to give a penetrating whistle. The song is a cheery, oft-repeated run of four notes *chwee che chu chu*, the last two notes are lower. Elaenias sit upright on perches like flycatchers with their wings held slightly drooped.

SIMILAR SPECIES No other resident birds have the combination of a fine bill and wing bars. North American warblers are similar but most have further distinguishing features such as streaking or more colour in the plumage.

WHEN TO SEE IT Elaenias are present year round and are a widespread breeding bird.

WHERE TO SEE IT They can be found across the island wherever there are taller bushes among the scrub and favoured areas include the West End, Katouche Valley woodlands and scrub near ponds such as Caul's Pond, East End Pond, Rendezvous Bay Pond and Long Pond.

LOCAL NOTES Look out for their flycatching behaviour where they fly from a perch after passing insects. Elaenias appear to breed from March or earlier, through to at least July. The nest is a shallow cup of twigs built in bushes and trees.

YOUR NOTES

Gray kingbird
Tyrannus dominicensis

Chincherry

HOW TO IDENTIFY IT L 22 cm (9"). This is a medium-sized land bird, grey above and white below with a neat black mask, a strong black bill and a slightly notched tail. They perch out in the open from a vantage point to look for insects often catching them with an audible snap. They are spectacular fliers and will make steep flapping display flights or noisily and aggressively pursue much larger birds intruding into their territory.

WHEN TO SEE IT A widespread breeding species, birds are present year round. More birds arrive in the period September–October when parties of up to 10 birds congregate on overhead wires although it is not clear if these birds stay or pass through on migration.

WHERE TO SEE IT Kingbirds are familiar birds across the island and perch singly or in small groups on wires, fence tops or tree tops around open areas. Regular sites include the grassland around Wallblake airport entrance and at the West End.

LOCAL NOTES The loud call gives rise to the local name of chincherry. Kingbirds breed mainly from April to June and build an open nest of twigs in trees and shrubs.

YOUR NOTES

Barn swallow
Hirundo rustica

Gale bird

HOW TO IDENTIFY IT L 15 cm (6"). A small, slim bird with long thin wings and a long forked tail, adult swallows are dark blue above with a reddish throat and pale orange underparts. Immature birds have more washed-out throat markings and lack the long tail streamers. Swallows hawk for insects low over water and grassland flying smoothly but erratically making it difficult to track an individual bird or count the flock. After feeding, they often gather on wires forming long lines. The typical flight call is a short, bright ringing *vit, vit*.

SIMILAR SPECIES This is the only regular swallow or martin seen on the islands although a few species occur as scarce visitors on migration. Caribbean martin is larger, purple above and dark on the throat and breast contrasting with a white belly. Bank swallow is smaller, dark brown above and white below with a brown breast band. Tree swallow has clean white underparts, a slightly notched tip to the tail and the adults are blue-green above.

WHEN TO SEE IT A few can be seen in most months from August–May although the daily totals vary with a handful of birds one day and flocks of 100 or more the next.

WHERE TO SEE IT They are most often seen over ponds feeding on aquatic insects with East End Pond, Road Salt Pond and West End Pond often attracting large flocks. They also feed over grassland particularly at Wallblake airport. Migrating birds may rest overnight around some ponds.

YOUR NOTES

Pearly-eyed thrasher
Margarops fuscatus

LOCAL NAME
Catbird, mockingbird

HOW TO IDENTIFY IT L 22 cm (9"). This thrush-sized land bird is brown above and white below with heavily streaked underparts. Look out for its pearly white eye (iris), strong yellowish bill and bold white spots to the tip of its tail. The thrasher has a range of calls including two to three clear whistling notes that can penetrate the tropical woodlands and a few, harsher more raucous calls. They can be bold and approachable around houses yet surprisingly wary in woodlands, quickly slipping away from view.

SIMILAR SPECIES Its size and plumage patterns are distinctive. Northern mockingbird is a rare migrant that occurs in the September–November period. These are greyer, unmarked below and have white in the wing and white wing bars.

WHEN TO SEE IT This is a common breeding bird and can be seen easily in all months.

WHERE TO SEE IT Thrashers are more common in the damper, wooded areas such as Katouche Valley and around ponds such as East End Pond and Blackgarden Bay Pond. They are also found around gardens with trees particularly those surrounded by scrub. In areas of housing they will fly up onto utility posts and buildings.

LOCAL NOTES Feeds on a variety of foods but will also attack the nests of other species such as bananaquits to take eggs and young chicks. Its shrill notes give rise to its local name of catbird. It breeds from March or earlier, through to at least July.

YOUR NOTES

Yellow warbler
Dendroica petechia

HOW TO IDENTIFY IT L 12 cm (5"). A small, colourful land bird, it is yellow-green above and brilliant yellow on all its underparts except for some faint reddish streaking on its breast and flanks. Younger birds are paler and greyer. They flit through bushes actively searching for prey, flicking their tail and occasionally dropping to feed on the ground. Its song is an attractive series of clear notes and the call a sharp *tzit*.

SIMILAR SPECIES Some North American warblers are similar with prairie warbler the most regular of these occurring from September–April. This has a more marked face pattern with a darker semi-circle under the eye, a dark spot on the side of the neck and streaking on the flanks. This streaking is black and heavy on adults and is a key feature. The back is also darker with faint, rufous streaks. A few migrant warblers occur more rarely (see list for species) and should be checked with a field guide.

WHEN TO SEE IT This is the only breeding warbler and is present year round. They are easier to see in the breeding season from April–July.

WHERE TO SEE IT A few can be found in mangroves, scrub and gardens across the island and on Prickly Pear East and Scrub Island. A walk around wetlands such as Little Harbour Ponds, Caul's Pond, Road Salt Pond and scrub such as at West End Point and around Junk's Hole Bay should provide sightings.

LOCAL NOTES The Caribbean population is considered a sub-species and is known as golden warbler *Dendroica petechia petechia*. Yellow warblers 'glow' in early morning light.

YOUR NOTES

Prairie warbler
Dendroica discolor

HOW TO IDENTIFY IT   L 12 cm (5"). A small migratory wood-warbler from North America, it is green above and yellow below. Adult birds have bold black streaks on the sides of their breast and flanks and these are paler in first year birds. With close views a dark crescent below the eye and a dark spot on the side of the breast are good features.

SIMILAR SPECIES   See text for yellow warbler. This is the most regular of the migrant wood-warblers although numbers can vary from year to year. There are several similar species and reference to a field guide is advised.

WHEN TO SEE IT   A non-breeding visitor, a few can be found from October–April with numbers varying year on year.

WHERE TO SEE IT   Prairie warblers are found in mangroves and woodland where it feeds among the tops of bushes and trees. With searching a few can be seen in suitable habitat across the island with Katouche Valley and Little Harbour Ponds the best sites.

YOUR NOTES

Blackpoll warbler
Dendroica striata

HOW TO IDENTIFY IT   L 13 cm (5.5"). Adult birds are greyish green above with streaks on the back, pale below with dark streaks on the breast sides and have a thin, dark line through the eye (females), two white wing bars and pale legs. Adult males have a bold black cap, white cheeks and a thin black stripe from the bill down the side of the throat. First-winter birds are more typical on Anguilla, are similar to adults but are more yellow on the throat and breast.

SIMILAR SPECIES   A number of similar warblers from North America could occur but only northern parula and black-and-white warbler are currently regular and both are distinctive. Parulas have two small wing bars but are pale grey on the head and wings, have white arcs above and below the eye and a clean yellow throat and breast with red patches. Black-and-white warblers are striking with a bold white eyebrow with a dark stripe above this on the sides of the crown, a streaked black-and-white back, two strong white wing bars and dark streaks on white underparts.

WHEN TO SEE IT   A few arrive each year in October with larger numbers grounded on occasion by storm systems. Most birds appear to move on although a few are likely to stay on until March.

WHERE TO SEE IT   If a large arrival occurs, birds can be found in trees and bushes across the island with Crocus Bay woodlands, Little Harbour Ponds and Road Salt Pond the most favoured areas.

LOCAL NOTES   Populations of a number of North American land birds winter in part within the Caribbean region or pass through on migration to Central and South America. Blackpoll warblers appear to be a passage bird on Anguilla. Anguilla's importance for this and other migratory warblers is still unclear although more information is now being gathered.

YOUR NOTES

Northern waterthrush
Seiurus novaboracensis

HOW TO IDENTIFY IT L 12 cm (5"). This is a skulking bird more often heard among mangroves and muddy edges of ponds. They are dark brown above and pale below with dense dark streaks on their breast and flanks, and fine dense streaks on the sides of the neck and throat. They have a fine bill, a prominent pale stripe above the eye and an explosive *spwik* call.

SIMILAR SPECIES It is always worth checking a field guide to confirm identification of warblers. Ovenbird is similar but is a rare visitor and is more olive-brown above, lacks the white eyebrow and has an orange crown with black borders.

WHEN TO SEE IT This is a scarce non-breeding visitor from October–April.

WHERE TO SEE IT Only found among mangroves to date feeding by muddy edges of ponds. It could occur on several ponds but all recent records are from Little Harbour Ponds, Blowing Point Pond and West End by-the-Sea Pond. Individual birds will spend several weeks in the same small area.

YOUR NOTES

Bananaquit
Coereba flaveola

Yellowbird

HOW TO IDENTIFY IT L 10 cm (4"). A popular and familiar bird across the Caribbean, their plumage is a colourful combination of dark grey upperparts contrasting with a sulphur yellow breast and a white belly and undertail. The plumage pattern is finished off with a dark throat, a white stripe above the eye, white wing spot and a yellow rump. Young birds are less bright and have a paler throat and a yellow eyestripe. Their short down-curved bill is used to take flower nectar often by making a hole in the base of larger flowers. These are noisy birds and will search restlessly through bushes and trees flicking their wings. Their typical calls include a sharp single or double *zit*. The song is often repeated several times and is a hurried drawn out jangle of notes with a final flourish; imagine the sound of sucking in through pursed lips. They build untidy round nests of grasses and will also build additional nests for roosting.

WHEN TO SEE IT This is a resident breeding bird and the commonest small land bird on the islands. The main breeding season is March–June but they will breed at other times following periods of heavy rain.

WHERE TO SEE IT Bananaquits are seen in all areas with trees and scrub and are hard to miss! Large concentrations occur in Katouche Valley and around Road Salt Pond for example. They will readily visit sugar feeders in gardens.

LOCAL NOTES Large groups will gather around trees such as white cedar when they are in flower. At other times they will feed on a variety of food including small insects. Look out for noisy chases through the scrub as birds pursue each other and adults fanning their broad tails flashing white tips.

YOUR NOTES

Male: right
Female: below

Black-faced grassquit
Tiaris bicolor

LOCAL NAME
Sparrow

HOW TO IDENTIFY IT L 11.5 cm (4.5"). These are small compact birds, largely olive-brown above and pale grey below with greyish heads. Only the adult males have a black face and they also have black underparts. The bill is small, grey and stout and with close views looks pointed. The male's song is a short, fast buzzing trill often followed by a single whistled note before repeating the song. The call is a soft *tsip*. Grassquits search for seeds on the ground and are often found in noisy flocks.

SIMILAR SPECIES The male bullfinch is all black with red on the chin and undertail. Female bullfinches are very similar to female grassquits but have blacker, more conical bills, have a distinctive orange undertail and are less grey on the face.

WHEN TO SEE IT A common resident and breeding bird it can be found across the island and the larger offshore islands.

WHERE TO SEE IT A walk through any area of open scrub or grassland should provide you with a few sightings and grassquits will readily visit gardens. Regular areas include the scrub alongside ponds such as the east side of Road Salt Pond and around Caul's Pond.

LOCAL NOTES Grassquits are known to breed year round although the main breeding season on Anguilla is March–August. The males display includes short flights with rapidly quivering wings.

YOUR NOTES

Male: right
Female: below

Lesser Antillean bullfinch
Loxigilla noctis

HOW TO IDENTIFY IT   L 14 cm (5.5"). A small bird, males are all black with a square red patch on the throat and undertail and a red spot above the eye although these are only visible with close views. Females are olive-brown above, including the head and grey-brown below with an orange-brown undertail. Bullfinches have short, black, conical bills adapted to dealing with a variety of seeds and fruits. Their calls include a short, high-pitched trill.

SIMILAR SPECIES   See text under black-faced grassquit.

WHEN TO SEE IT   A species present year round breeding in small numbers across the island. They breed from March–August at lower densities than grassquits.

WHERE TO SEE IT   Bullfinches are typically found where there are mature trees and bushes but can be inconspicuous and hard to find. They will return to a good food source such as tree buds. They will visit mature gardens and are often found around Caul's Pond, Cove Bay Pond, Little Harbour Ponds and woodland at the West End.

LOCAL NOTES   This species is found throughout the Lesser Antilles and appears to be more tame and approachable on some of the other islands in its range.

YOUR NOTES

Other land birds

Rock dove (feral)
Columba livia
A scarce resident bird, these town pigeons are not native to the region and were first seen on the island in 2000. Typically seen perched on buildings, there is a large breeding population on nearby St Martin.

Scaly-naped pigeon
Columba squamosa
A rare visitor. A single record of a bird seen in the grounds of the Mallihouana Hotel in March 1995.

White-crowned pigeon
Columba leucocephala
A rare visitor following a local extinction in the 1960s through hunting and habitat loss. A single bird was seen around Blackgarden Bay Pond in August 2005.

Eurasian collared dove
Streptopelia decaocto
A rare visitor this is an introduced species to the region from Eurasia. It is common on nearby St Martin and Saba although there are only two sightings on Anguilla to date at Old Ta Road in September 1995 and by Shoal Bay East in April 2005.

White-winged dove
Zenaida asiatica
A scarce visitor, several birds are now established following the arrival of the first birds in 2000. They are most often seen perched on wires and favoured areas include the area around Little Bay and East End village.

Yellow-billed cuckoo
Coccyzus americanus
A scarce visitor, varying numbers can be seen across the island in the period September–November. This species is more obvious than mangrove cuckoo when present and following thunderstorms in October 2006 at least 10 birds were seen with the areas around Long Pond and Road Salt Pond providing most sightings.

Common nighthawk
Chordeiles minor
A scarce visitor, this species is probably under-recorded as it is very difficult to separate from Antillean nighthawk. In October 2006 over 20 birds arrived when migrating North American land birds were grounded by heavy thunderstorms.

Antillean nighthawk
Chordeiles gundlachii
A breeding visitor in small numbers, this species can be seen from April–October. They are seen at dusk and dawn flying with light, stiff wing beats chasing insects. The males display call is far-carrying and gives rise to a local name of *Gimme-me-bit* in some parts of the Caribbean. Look for them inland from Limestone Bay, the headland at Sandy Ground (Road Point) and inland from Little Harbour.

Chimney swift
Chaetura pelagica
A rare visitor, a single bird was seen with barn swallows over Mead's Bay in October 2006.

Purple-throated carib
Eulampis jugularis
A rare visitor, there is a single record of a bird at East End village in October 1996. Anguilla is near the northern edge of the range for this species and the nearest population is on Saba.

Antillean crested hummingbird
Orthorynchus cristatus
A scarce resident and probable rare breeding bird with one to three birds seen in recent years in the Katouche Valley and the Little Harbour area. This tiny bird disappeared from the island in the wake of Hurricane Luis in 1995 and it was December 2001 before a male was again seen.

Caribbean martin
Progne dominicensis
A rare visitor to headlands and over ponds with two sightings in recent years at Cove Bay Pond in April 2000 and over Island Harbour in August 2005. All reports of this species have involved brief views suggesting it is overlooked and a scarce visitor with records most years.

Tree swallow
Tachycineta bicolor
A rare visitor, there are three reports involving one bird near The Valley in December 1993, four at Long Pond and one over Sombrero, both in November 1999.

Bank swallow
Riparia riparia
A rare visitor but likely to be over-looked among the many flocks of barn swallows passing over the

island. Sightings involve one near The Valley in December 1993, two over Badcox Pond in January 2000 and one over Mead's Bay in October 2006.

Northern mockingbird
Mimus polyglottos
A rare visitor from North America, the only recent sighting is one at The Farringtons in November 1999.

Northern parula
Parula americana
A scarce visitor, a few are seen most years in the period October–April. They have been found in a number of areas including Katouche Valley, Little Harbour Ponds and on Sombrero.

Magnolia warbler
Dendroica magnolia
A rare visitor, there is one record of a single bird in Katouche Valley in February 2003.

Black-throated blue warbler
Dendroica caerulescens
A rare visitor, there is a single report of a male on Sombrero in November 1999.

Yellow-throated warbler
Dendroica dominica
A rare visitor, one was seen in The Valley in January 1996.

Yellow-rumped warbler
Dendroica coronata
A rare visitor to wooded areas around ponds, there are three records in the period December–April with sightings at Caul's Pond, Cove Bay Pond and the Katouche Valley.

Black-throated green warbler
Dendroica virens
A rare visitor, there are three records involving a male on Sombrero in November 1999, one in the Katouche Valley in January 2004 and a male at Malliouhana Hotel in April 2006.

Black-and-white warbler
Mniotilta varia
A scarce visitor, small numbers can be found in the Crocus Bay and Katouche Valley woodlands in the period October–February. Look for these strikingly marked, heavily streaked warblers low down in trees where they forage on trunks and along branches.

Prothonatary warbler
Protonotaria citria
A rare visitor, there was a single bird in Katouche Valley in February 2003.

Ovenbird
Seiurus aurocapillus
A rare visitor, a single bird was seen on Sombrero in November 1999.

Swainson's warbler
Limnothylypis swainsonii
A rare visitor, a single bird was present in Katouche Valley in January 2004.

Common yellowthroat
Geothlypis trichas
A rare visitor, four were found on Sombrero in November 1999 and a male was by Caul's Pond in April 2000.

Hooded warbler
Wilsonia citrina
A rare visitor with single birds seen at Little Harbour Ponds in March 1996, on Sombrero in November 1999 and in the Katouche Valley in February 2003.

Antillean euphonia
Euphonia musica
A rare visitor from the Lesser Antilles that has been reported in the past but with only a single recent record of one by Gull Pond in February 2000.

Scarlet tanager
Piranga olivacea
A rare visitor from North America with a single bird at Sandy Ground in November 1999. Reports of unidentified red birds (males) suggest it is under-recorded on the island.

Shiny cowbird
Molothrus bonariensis
A rare visitor, a male was seen near Sandy Hill in April 2000.

House sparrow
Passer domesticus
A scarce resident, a small population has become established in Island Harbour since 2002 with over 30 birds seen recently. This species is not native to the region but has established populations on several islands including St Martin and Saba.

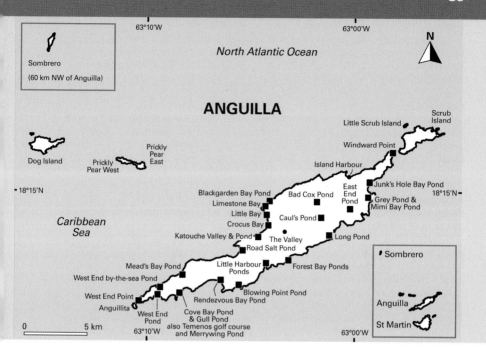

Birdwatching sites

How to use this section

The featured sites in this section include most of the sites mentioned elsewhere in the guide. A number of other sites are described at the end of this section including the outer islands. The following notes are intended to help you make the most of your visits and help you enjoy your time birdwatching on Anguilla.

General notes

Early morning and late evening are the best time to see birds. In general, the eastern side of sites provides better light in the morning and the western side in the evening. The temperature is often cooler at each end of the day. Temperatures can rise quickly, however, and care should be taken over sun protection and the provision of drinks.

Each habitat presents a different experience and conditions. Limestone areas can include sharp, heavily fissured rock, particularly near the coast and require suitable footwear. In these areas, paths are often overgrown with low thorn bushes and cacti, again requiring suitable footwear and care. Insect bites are possible particularly around wetlands and suitable protection should be used.

Directions

All site directions are from the Anguilla National Trust (ANT) building in The Valley and distances are approximate and given in miles. We recommend picking up a map of the island provided free by the Anguilla Tourist Board and widely available at hotels and a range of other outlets. The sites in this section run clockwise around the island following the coast starting with Bad Cox Pond in the east.

There is a limited bus service around the island and several sites are close together and can easily be visited on foot or by bicycle. Car hire and taxis are also available and the directions follow the road network as there are few walking trails currently on the island.

The road network and signage continues to be upgraded and sites are easily reached by car unless difficulties are mentioned in the text. Some routes can be temporarily inaccessible due to flooding after heavy rain and some of the limestone rock roads can become heavily fissured during storms making access difficult by vehicles.

Maps and access
The maps have been compiled to highlight the main access routes, viewing points and information on the main natural features. The majority of the sites featured are in private ownership and we have only featured viewing points that are from public tracks. Most landowners are happy to allow access for birdwatching but please respect any signs restricting access. If access is unclear or you would like further information, please contact the ANT. ☀ = recommended viewing point.

Viewing the sites
This section includes a description of the site and some of the plants and other wildlife to be found there. It also includes information on how to get the most out of your visit.

Birds found at the site
The lists of bird species provide an indication of species typically found at the site year round or at different times of year. These lists are indicative and all the species may not be present on each visit and some species may be found outside the period for which they are listed. The bird lists for seasonal ponds that dry out only apply when the pond is holding water. As a rule the more time spent at a site the more bird species will be seen. For more up-to-date information on scarce or rare birds please contact the ANT.

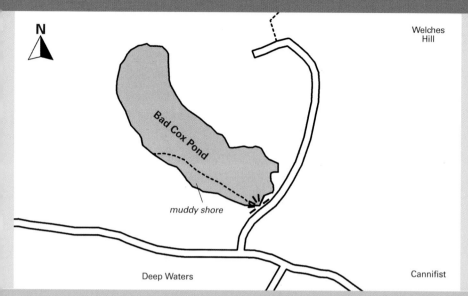

N

Welches
Hill

Bad Cox Pond

muddy shore

Deep Waters

Cannifist

Bad Cox Pond

ACCESS From the ANT office leave in the direction of the traffic lights. Turn left at the lights and follow the road for 1.9 miles until the roundabout at the Shoal Bay East turn. Take the right turn onto the Deep Waters Road and Bad Cox Pond is 0.6 miles along this road on the left. Turn left on the track on the brow of the hill with a line of utility poles. A viewing point is on the track when you meet the pond on your left after 0.2 miles. The pond is privately owned.

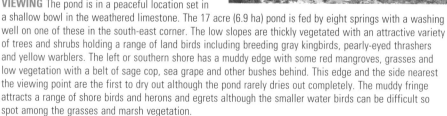

VIEWING The pond is in a peaceful location set in a shallow bowl in the weathered limestone. The 17 acre (6.9 ha) pond is fed by eight springs with a washing well on one of these in the south-east corner. The low slopes are thickly vegetated with an attractive variety of trees and shrubs holding a range of land birds including breeding gray kingbirds, pearly-eyed thrashers and yellow warblers. The left or southern shore has a muddy edge with some red mangroves, grasses and low vegetation with a belt of sage cop, sea grape and other bushes behind. This edge and the side nearest the viewing point are the first to dry out although the pond rarely dries out completely. The muddy fringe attracts a range of shore birds and herons and egrets although the smaller water birds can be difficult so spot among the grasses and marsh vegetation.

The pond often attracts feeding brown pelicans and is an important area for dabbling ducks in the period October–March. The two main species are blue-winged teals and white-cheeked pintails and they often roost on the secluded far bank. Wetland bird counts on this pond have discovered the occasional green-winged teal and American wigeon in recent years.

WHAT TO SEE **YEAR ROUND**: green heron, white-cheeked pintail, common moorhen, killdeer, black-necked stilt, gray kingbird. **MAY–AUGUST**: snowy egret. **SEPTEMBER–APRIL**: great egret, great blue heron, green-winged teal, blue-winged teal, American wigeon, spotted sandpiper, solitary sandpiper. **RARITIES**: northern shoveler, bank swallow.

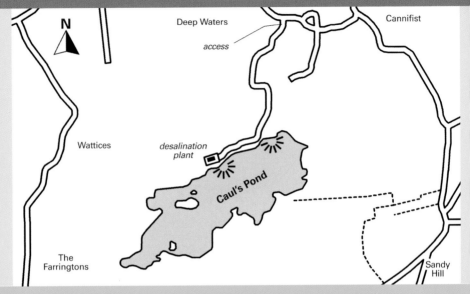

Caul's Pond

ACCESS From the ANT office leave in the direction of the traffic lights. Turn left at the lights and follow the road for 1.9 miles until the roundabout at the Shoal Bay East turn. Take the right turn onto the Deep Waters Road and after 0.9 miles turn off right onto the track just before the West Indies Concrete Company works on your left. Follow the road and bear right at the first opportunity. Pass through the houses and after the last building the track turns sharply left and then sharply right. At this point the track drops down a hill and becomes heavily fissured in places. If in a car, park here and walk down to the pond. At the pond edge the track turns right and runs for 0.3 miles to a disused water desalination plant. The pond is in public and private ownership.

VIEWING The scrub on the track to Caul's Pond is rich in trees, bushes and flowers and attracts most of the island's land birds as well as insects and lizards. Limestone rocks and scrub drop down to the northern, eastern and southern shores and overhang the pond in places. Mangroves and buttonwood grow on the muddy shores. The pond covers 91 acres (36.8 ha) when full and is considered one of the most natural and unaltered ponds of its kind in the Lesser Antilles. The pond is fed by freshwater springs and always retains a large body of water. This pond was noted for large gatherings of water birds until a mosquito control programme in 1987 resulted in a bird kill and affected the ecology of the pond. Twenty years on, large flocks of egrets and shore birds suggest it is returning to its former glory.

The limestone rocks at the bottom of the track provide the first viewing point for the eastern side of the pond. You can sit on a limestone platform here almost at water level. The bay on your left can be good for ducks and coots. The eastern and central parts of the pond are deep enough for diving ducks and pelicans to feed. The far bank is less disturbed and attracts herons, egrets and further groups of ducks. The dead trees by the water can attract common moorhens and green herons and the bushes can be good for resident land birds and occasional migratory warblers. Moving west, the next viewing point is by a fenced-off desalination plant. Check the trees on the opposite bank for roosts of egrets and the shallow muddy fringes for shore birds. It is difficult to get close to the western edge so a telescope is essential. Look out for ospreys as they perch in the dead trees on the western edge. The shallower western side dries out in drier periods when it becomes an important feeding area for shore birds.

WHAT TO SEE **YEAR ROUND**: snowy egret, green heron, white-cheeked pintail, common moorhen, green-throated carib, Caribbean elaenia, yellow warbler. **MAY–AUGUST**: laughing gull. **SEPTEMBER–APRIL**: pied-billed grebe, great egret, ring-necked duck, lesser scaup, ruddy duck, osprey, American coot, killdeer, whimbrel. **RARITIES**: northern shoveler, yellow-rumped warbler, common yellowthroat.

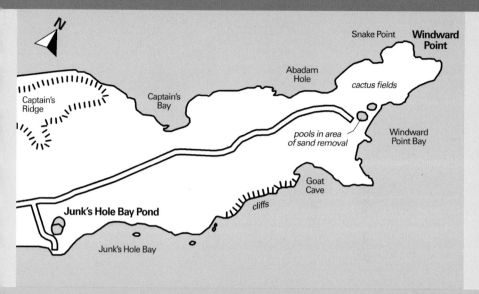

Windward Point and Junk's Hole Bay Pond

ACCESS Leave the ANT building in the direction of the traffic lights and go straight across following all road signs to East End Village. Continue along the main road around East End Pond and stay on the main road for a further 1.1 miles until as you reach a small incline you see a supermarket on the right. Take the track immediately to the right of the supermarket. This quickly becomes a sandy track and is rutted in places. Follow the track towards Junk's Hole Bay and go straight over where the tracks cross. At the second fork, turn right following the sign for Palm Grove Bar. Park outside Palm Grove Bar and walk around the front of the bar and approach the pond from the beach.

For Windward Point, return to the main track and turn right. Follow this as far as possible keeping straight on. Park near the sand extraction area and follow the beach towards the beacon on top of the point. The pond and area are privately owned.

VIEWING This whole area is among the most remote and least visited parts of the island and has special qualities as a result, although the north coast is now seeing new development. The area around Junk's Hole is an attractive landscape of low windswept scrub punctuated with frangipani and tall doodle doo cacti that sweeps down to Junk's Hole and Savannah Bays. A few land birds move through the scrub including mangrove cuckoos. The pond is small and shallow and can hold a surprising variety of water birds. If you approach through the dunes there is a gap to view the pond through the sea grape bushes, mangrove and buttonwood. Shore birds will roost on the low walls from former salt workings, providing close views. The area is good for ground lizards and the beach attracts a few shore birds and terns offshore.

Towards the Point, the shallow, seasonal pools that form the sand quarry attract egrets, a variety of shore birds in small numbers and are particularly good for roosting terns either on migration or from the nearby breeding colonies on Scrub Island. All the grey terns occur here and it has become the easiest place to see roseate terns.

The walk to the point is rocky but worth it as it takes you through a magical field of Turk's head cactus, the largest population on the island. The Point itself offers good views across to Scrub Island and occasional close views of seabirds including tropicbirds. Swallows and martins moving across the island tend to circle here before moving off west.

WHAT TO SEE YEAR ROUND: red-billed tropicbird, brown booby, brown pelican, royal tern, common ground-dove, mangrove cuckoo, green-throated carib. **MAY–AUGUST**: Sandwich tern, roseate tern, least tern. **SEPTEMBER–APRIL**: spotted sandpiper, ruddy turnstone, red knot, sanderling, barn swallow.

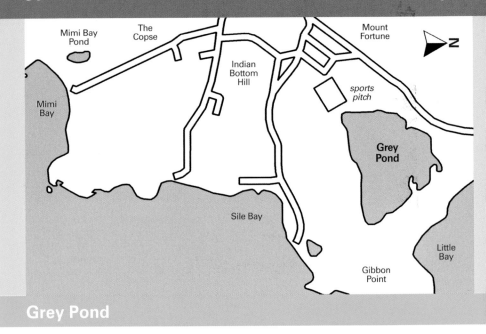

Grey Pond

ACCESS Leave the ANT in the direction of the traffic lights and continue straight on (east) following all road signs to East End Village. At the sharp left hand bend at the end of the pond turn right (second right after the museum) along the tarmac road into the village. After 0.25 miles take the first right. Take the left fork after 100 metres. In a further 100 metres you arrive at a fenced sports field. Walk on the outside of the fence on the left hand side down to the pond. The track to the pond edge is in line with the fence but is overgrown and the path winds round to the left hand palm tree. This area is privately owned.

VIEWING The 36 acre (14.5 ha) pond is surrounded by low sand dunes from Savannah Bay and has high landscape value. Fresh breezes from the prevailing winds blow over the pond giving it a wilder feel. The pond quickly reflects the mood of the weather and can look grey and desolate in stormy weather. The access track takes you to the edge of the pond where muddy edges, covered in part by stands of salt meadow plants such as spike rush and pondweed, attract a range of shore birds. The large pond has a mix of shallow rock pavement and softer muddy shores attracting ducks and roosting shore birds. Plovers, willets and least terns will breed here and can be viewed with a telescope. The pond is surrounded by low scrub on the rock pavement and the steeper slopes on the north side can provide elevated but more distant views.

WHAT TO SEE YEAR ROUND: white-cheeked pintail, black-necked stilt. **APRIL–AUGUST**: willet, laughing gull, least tern. **SEPTEMBER–MARCH**: black-bellied plover, semi-palmated plover, killdeer, whimbrel, stilt sandpiper. **RARITIES**: northern pintail.

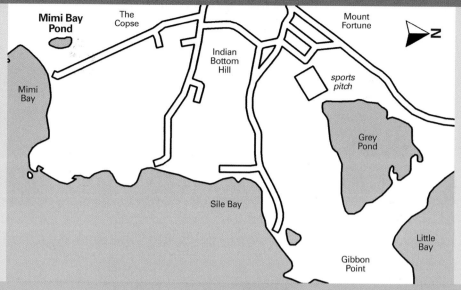

Mimi Bay Pond

ACCESS Leave the ANT office in the direction of the traffic lights and continue straight on (east) following all road signs to East End Village. On entering the village take the first right after the museum (Brother David's Drive). Take the first track on the left hand side after 100 metres. Take the first right hand fork and continue straight along the track taking the next fork to the right onto a grassy pathway. Drive straight over the brow of a hill where the pond can be glimpsed on the right hand side amid the scrub. Park here by some large rocks on the left hand side and walk through the track to

the right in the sea grape and buttonwood bushes to access the stone pavement which provides a viewing platform. This area is privately owned.

VIEWING The access track runs for 100 metres across limestone rock and pockets of sand. The scrub on the rocks includes frangipani and Turk's head cactus. The sandy pockets hold sea grape bushes and cedars attracting a range of land birds, some large ground lizards and hermit crabs. Find a vantage point on the rock platform above the pond. Be careful to avoid the low spiky shrubs in this area. The pond has filled a small hollow in the rock and has developed a thick border of red mangroves and buttonwood trees among dead tree stems from recent hurricane damage. Ducks and shore birds hide and roost in this cover or perch on the dead stems. The surrounding scrub has some lovely wind sculptured bushes that sweep down to the secluded Mimi Bay and beyond with attractive views across to St Martin.

WHAT TO SEE **YEAR ROUND**: black-necked stilt, green-throated carib, Caribbean elaenia.
MAY–AUGUST: white-cheeked pintail, laughing gull. **SEPTEMBER–APRIL**: blue-winged teal, merlin, lesser yellowlegs.
RARITIES: northern shoveler.

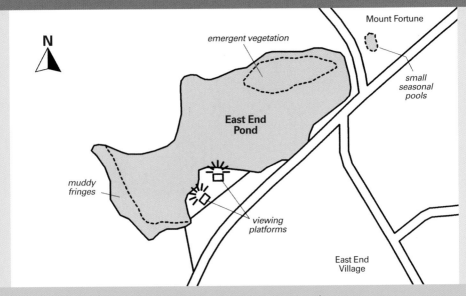

East End Pond

ACCESS Before you leave the ANT office ask for the keys to the gate to the viewing platforms. Leave the ANT building to the right towards the traffic lights. Go straight across and stay on the main road as it goes along the long road to Sandy Hill. When you reach a double roundabout follow signs to the right for East End Village. The pond is on your left as you reach the village. Pull onto the parking area in front of the fence. The pond is in public ownership.

VIEWING This shallow, seasonal pond of 13 acres (5.3 ha) is fed by springs and has a mix of open water, surface vegetation and muddy fringes. This combination makes it one of the best birdwatching sites on the island and it is at its best when high water levels are beginning to drop exposing muddy areas. The pond's small size enables close views of most of the island's wetland and land birds. The site is now a protected area managed by the ANT and they have erected two raised and covered viewing platforms. The pond is hemmed in by the road on two sides, provides a community resource for the school and village and is now more secure than in the past when an area of the southern edge was filled in or tipped on. The rear of the pond has a limestone edge and a thick cover of scrub providing perches for egrets and herons and safe resting areas for water birds that are still close enough to provide good views. The west side is shallower and dries out first providing feeding areas for a wide range of shore birds. The east side is deeper and is one of the few wetland areas with emergent vegetation favoured by blue-winged teals, ruddy ducks, common moorhens and coots. The flocks of ducks have held green-winged teals and ring-necked ducks in recent years. Very few birds are present when the pond is completely dry. Across the road at the eastern end of the pond are a few small pools and a marshy area worth checking with green herons and Wilson's snipe recorded here.

WHAT TO SEE **YEAR ROUND**: great egret, snowy egret, green heron, white-cheeked pintail, common moorhen, American coot, Caribbean coot, black-necked stilt, common ground-dove, gray kingbird. **MAY–AUGUST**: killdeer, willet, laughing gull. **SEPTEMBER–APRIL**: pied-billed grebe, tri-coloured heron, green-winged teal, blue-winged teal, ring-necked duck, ruddy duck, merlin, peregrine, black-bellied plover, whimbrel, Wilson's snipe, barn swallow. **RARITIES**: snow goose, West Indian whistling duck, northern shoveler.

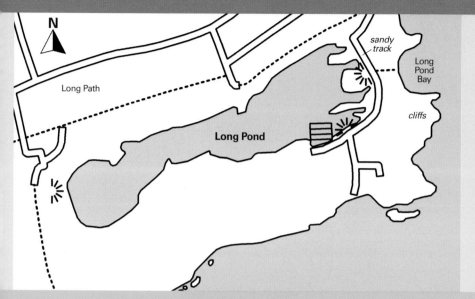

Long Pond

ACCESS Landward end: Leave the ANT building and turn right towards the traffic lights in The Valley. Go straight across and follow the road for 2.5 miles joining Long Path Road. Just before a slight incline turn right in front of a two storey house. Follow the track down for 0.5 miles (it turns right and then left at the bottom). The pond is viewable on the left of the track. **Seaward end**: Return to the main road and turn right following Long Path Road for a further 1.8 miles. After the petrol station on your left, take the first main track on your right, just before the road bends round to the left. Follow the track down past the houses and if in a car, park at the end. Walk down the track passing through stands of sea grape bushes. The track continues into the dunes with the pond visible on your right and Long Pond Bay and beach on your left. This site is privately owned.

VIEWING This site is a large, shallow coastal lagoon of 50 acres (20.2 ha) when full, separated from the sea by low dunes and an open sandy area. The sea rarely breaches the sand dunes, but when full after heavy rains the pond overflows back into the sea. In dry periods, the pond begins to dry out but rarely to more than half its extent. When water levels are low the remains of stone dykes from past salt production and a former shrimp farm can be clearly seen from the seaward end. Access to both ends of the pond requires a short walk and the ground can be muddy on the landward end if you leave the access road. The northern and southern shores are coralline limestone covered in low scrub.

The seaward end has a mix of sand dune and salt meadow plants with pioneering buttonwood bushes. Look out for seashore papsalum grass, sea bean, sea lavender and seaside spurge. On the landward side torrey, a salicornia, grows on the mudflats along with pondweed.

The site can be rich in prey such as brine shrimps and attracts some of Anguilla's largest flocks of shore birds at peak times and feeding flocks can be seen at both ends of the pond. The light and distance from viewing points can make identification difficult at times and a telescope is recommended for close views. Feeding birds can easily be disturbed so try to find cover and not to get too close. This is particularly important on the seaward end where a broad sandy area is one of the best places remaining on the mainland for breeding least terns and shore birds such as Wilson's and snowy plovers.

The narrow beaches and low rocky outcrops of Long Pond Bay attract feeding and roosting shore birds.

WHAT TO SEE YEAR ROUND: snowy plover, Wilson's plover, common ground-dove. **MAY–AUGUST**: willet, laughing gull, roseate tern, least tern. **SEPTEMBER–APRIL**: osprey, black-bellied plover, semi-palmated plover, lesser yellowlegs, ruddy turnstone, red knot, semi-palmated sandpiper, least sandpiper, stilt sandpiper. **RARITIES**: little blue heron, piping plover.

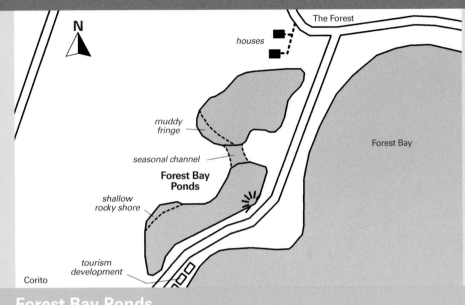

Forest Bay Ponds

ACCESS Leave the ANT building and follow the main road to the west of the Island. At the traffic lights near the western end of the airport, turn left into Jeremiah Gumbs highway. After 0.5 miles, turn left with the airport on your left. At the end of the airport, follow the road round to the right. After 0.3 miles, take the left fork and follow this down to the bay. At the dunes, turn right and the ponds are on your right. The ponds are privately owned.

VIEWING The ponds are a string of three pond basins that form a single 6 acres (2.4 ha) pond when full. The pond lies in between a shallow band of coastal dunes and a scrub covered limestone slope.

The vegetation is recovering from recent hurricane damage and the dunes contain a range of bushes including tamarind and sea grape.

The southern pond is visible from a couple of points off the tarmac track through the dunes and allows close views of egrets, herons and shore birds. The northern pond retains water run-off and currently is only visible from a viewpoint reached a short distance back up the approach road. This has a muddy area at the rear, attracting egrets, breeding common moorhens and stilts. A range of shore birds will feed here although views are more distant. The ponds are surrounded by all the mangrove species and at times are used by roosting egrets. Belted kingfishers fish around these ponds and in the surf in Forest Bay where ruddy turnstones can be found on the beach.

WHAT TO SEE YEAR ROUND: snowy egret, common moorhen, black-necked stilt, Caribbean elaenia, yellow warbler. **MAY–AUGUST**: willet. **SEPTEMBER–APRIL**: brown pelican, great blue heron, great egret, greater and lesser yellowlegs, ruddy turnstone, belted kingfisher.

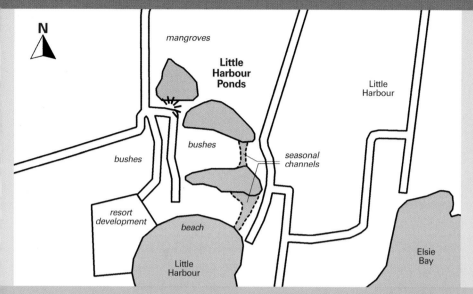

Little Harbour Ponds

ACCESS Leave the ANT building and follow the main road to the west of the Island. At the traffic lights near the western end of the airport, turn left onto Jeremiah Gumbs Highway. After c0.8 miles at the bottom of a hill, look out for a track off to the left. This quickly becomes a rock track and where it forks take the left track. The ponds are on your left. This area is privately owned.

VIEWING This is one of the best remaining areas of mangroves on Anguilla growing around a series of small ponds at the base of a limestone hill. The ponds are seasonal, fed by rainfall and rarely dry out fully. They also receive occasional storm surges and the ponds are still recovering from the hurricanes in the 1990s. Several years on, the vegetation is now becoming dense providing safe roosting and feeding areas for water birds although making it increasingly difficult to find viewing points. There is a good opportunity to view the large pond on your left from the first telegraph pole. It is worth spending some time here as birds such as pied-billed grebes, herons, egrets and ducks will move in and out of cover. The bushes between here and the beach are among the best places to find migratory warblers including the blackpoll warbler, prairie warbler and northern waterthrush and when water levels are high, sora rails are often present. The large pond nearest the bay can be viewed off the beach from either end. The coastal creeks good for yellow-crowned night-herons, green herons and belted kingfishers. The bay is very shallow and receives little tidal inflow attracting few birds. The rocks at its mouth occasionally attract American oystercatchers.

WHAT TO SEE **YEAR ROUND**: green heron, yellow-crowned night-heron, white-cheeked pintail, common moorhen, black-necked stilt, mangrove cuckoo, pearly-eyed thrasher, yellow warbler. **MAY–AUGUST**: willet. **SEPTEMBER–APRIL**: pied-billed grebe, great blue heron, great egret, tri-coloured heron, blue-winged teal, American wigeon, osprey, sora rail, whimbrel, Wilson's snipe, belted kingfisher, barn swallow, northern parula, prairie warbler, blackpoll warbler, northern waterthrush. **RARITIES**: little blue heron, hooded warbler.

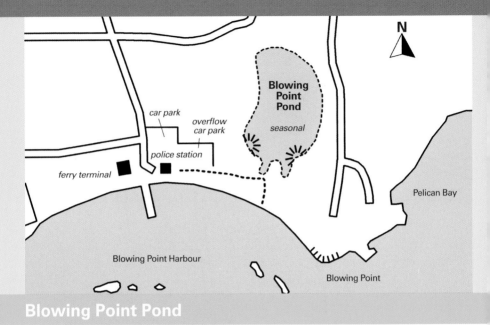

Blowing Point Pond

ACCESS Leave the ANT towards the roundabout and turn left and follow the main road west. At the traffic lights near the western end of the airport turn left into Jeremiah Gumbs highway. Continue to the end of the highway and then turn left towards the ferry terminal. From the car park immediately in front of the police building, turn left, and follow the track through the overflow car park until you reach the pond on your left. The pond is privately owned.

VIEWING This is a shallow seasonal pond of 47 acres (19 ha), and is quick to dry out. It has been neglected in the past but at times it can be one of the best ponds for shore birds on the island. The north edge has low stumps of dead mangroves that egrets, herons, whimbrels and lesser yellowlegs perch on. The vegetation around the pond is heavily grazed by goats, however, buttonwood and other bushes are now recovering at the rear. This area dries out first and is often very good for shore birds, particularly plovers, although a telescope is essential for good views. The pond is usually empty of birds when dry although a few black-bellied plovers can often still be found.

The mangroves and bushes on the seaward end provide cover, are very good for ground lizards, and with care you can find a viewing point to see egrets, herons and shore birds at close quarters. The creeks nearest the sea often have fish in them and can hold night-herons, sora rails, spotted sandpipers and warblers. Check the rocks offshore for American oystercatchers and terns.

WHAT TO SEE **YEAR ROUND**: royal tern, yellow warbler. **MAY–AUGUST**: Wilson's plover, willet, laughing gull. **SEPTEMBER–APRIL**: great egret, yellow-crowned night-heron, osprey, sora rail, black-bellied plover, semi-palmated plover, spotted sandpiper, whimbrel, ruddy turnstone, belted kingfisher, barn swallow, prairie warbler, northern waterthrush. **RARITIES**: great black-backed gull.

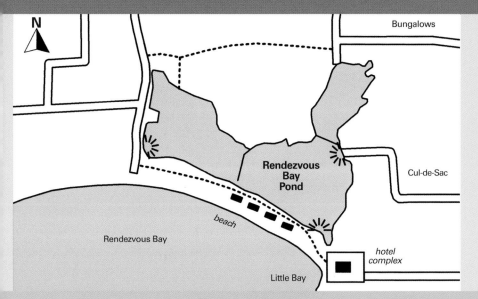

Rendezvous Bay Pond

ACCESS Eastern side: Leave the ANT building turning left to the roundabout and turn left to follow the main road to the west of the Island. At the first traffic lights near the western end of the airport, turn left onto Jeremiah Gumbs highway. Follow this to the end into Blowing Point village. Turn left and then first right signposted to Rendezvous Bay Hotel. When you arrive at the resort, walk to the bay and turn right travelling west along the beach. The pond is on your right hand side. For an alternative eastern viewing point on the pond when you reach the end of the Jeremiah Gumbs highway, travel straight across

and follow the road until you reach the pond. Western side: Returning to the turning in Blowing Point village turn left and follow the road until you reach the main road heading west across the island. Turn left and head west for 1.4 miles before turning right following signs for The Anguilla Great House. Continue to the end of the road and you will find the pond on your left. The pond is in public ownership.

VIEWING The pond is a large coastal lagoon covering 56 acres (22.6 ha) and rarely dries out completely. Walls remain from former salt production providing safe resting areas for a range of water birds. The pond is separated from the sea by a sand bar behind the beach on the southern shore and an access road on the west. Low, densely vegetated limestone rocks flank the northern and eastern side. Access is difficult away from the viewing points and a telescope is required to scan the far banks.

In most bays a few water birds will be present such as snowy egrets, white-cheeked pintails and spotted sandpipers. White-cheeked pintails, black-necked stilts and Wilson's plovers will breed along with least terns. Belted kingfishers occasionally visit the pond and the surrounding scrub is excellent for land birds including the American kestrel, Caribbean elaenia, yellow warbler and black-faced grassquit. The bay is one of the island's finest and attracts common seabirds such as the brown booby and royal tern.

WHAT TO SEE YEAR ROUND: white-cheeked pintail, American kestrel, Caribbean elaenia, yellow warbler, black-faced grassquit. MAY–AUGUST: Wilson's plover, laughing gull, least tern. SEPTEMBER–APRIL: snowy egret, lesser yellowlegs, spotted sandpiper.

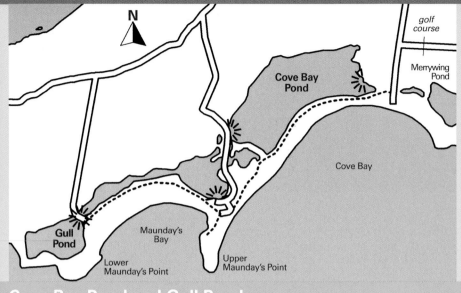

Cove Bay Pond and Gull Pond

ACCESS This long coastal lagoon can be reached from three access points at its eastern, central and western edges. **Eastern edge**: leave the ANT building to the roundabout and turn left heading out to the west of the island. Stay on the main road for a total of 6.4 miles until you reach a petrol station on your left. Turn left immediately after the petrol station and continue to the bottom of the track ending at the dunes with the pond on your right. **Central section**: return to the main road and turn left. After a further 0.8 miles follow signs to Cap Juluca resort taking the left hand turn down across a causeway into the resort. If travelling by car, park in public car park by the beach. **Western section**: return to the main road and turn left. Take the first left after 0.5 miles down Firefly Lane. Continue to the bottom and park along the causeway. The western end of Cove Pond is on your left and Gull Pond is on your right. The pond is in public ownership.

VIEWING These ponds form the largest coastal lagoon complex on the island and have the highest remaining dune systems although these were damaged in the last hurricane and sections have been developed for the Cap Juluca resort. The lagoon is now divided by two causeways, a central one to the resort and a second service access at the western end. The large eastern section is known as Cove Pond and covers 114 acres (46 ha). It is thought Merrywing Pond, now part of a new golf course complex, was formerly part of the same unit.

The eastern edge can be viewed from the end of the road or from a track through the dunes, with egrets and shore birds scattered around the edges of the pond. Cove Pond, although shallow, tends to maintain its water area and holds populations of fish which attract a few pelicans to feed. Cross the road to the east here and you can look over Merrywing Pond now modified and part of the new golf course complex.

The western side of Cove Pond and eastern side of Gull Pond have sand bars colonised by buttonwood and salt meadow plants. These areas hold breeding least terns, Wilson's and snowy plovers although there has been too much disturbance in recent years. The area section behind the resort is recovering from past hurricane damage and now has over-hanging vegetation including buttonwood and black mangroves and the resort has set out a short trail to help identify the plants. This area is a favourite haunt and roost site of ducks, stilts, egrets and herons and the path to the trail attracts a variety of land birds. The western side of Gull Pond, also known as Maunday's Bay Pond, is connected to the main lagoon by culverts under the service road. This area attracts egrets and shore birds including Wilson's plovers to the rocky and sandy northern edges.

WHAT TO SEE **YEAR ROUND**: snowy egret, white-cheeked pintail, black-necked stilt, mangrove cuckoo. **MAY–AUGUST**: snowy plover, willet, Sandwich tern, least tern. **SEPTEMBER–APRIL**: sora rail, semi-palmated plover, Wilson's phalarope, laughing gull. **RARITIES**: greater flamingo, black-throated blue warbler, Antillean euphonia.

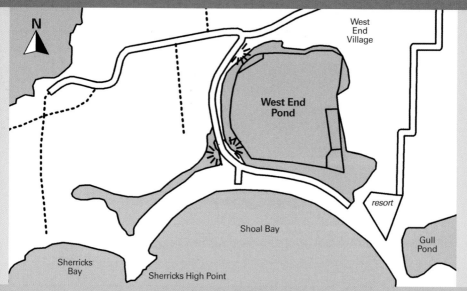

West End Pond

ACCESS Leave the ANT building turning left towards the roundabout. Follow the main road all the way to the west of the island, a journey of 8.1 miles. At the very end, a tarmac road bears off to the left forming a causeway across the pond. When you reach the resort buildings, park here and walk along the southern edge of the pond. This is a privately owned pond.

VIEWING This is the last of the big coastal lagoons running westwards from Rendezvous Bay. This shallow salt pond covers 41 acres (16.6 ha)

and retains most of the pond walls from salt working that continued until 1976. The salt pond is slightly deeper in the centre and tends to retain some water through all but the driest periods. The narrow western arm has a different feel and vegetation with overhanging mangrove and buttonwood providing cover for breeding ducks and stilts.

The main pond and the views from the causeway can provide some of the best views of shore birds on Anguilla. Ducks, terns and shore birds will rest on salt pond walls and least terns occasionally breed on them. When the pond is full, a variety of shore birds will feed close to the road providing excellent opportunities for identification and to compare their feeding methods. A feature of this pond is the large hatches of brine flies that attract gulls, shore birds and swallows.

The pond has areas of scrub around its eastern and western edges that hold a range of land birds. The dunes system is now largely developed and fronts onto Shoal Bay West, a good bay to watch for brown boobies and royal terns.

WHAT TO SEE YEAR ROUND: white-cheeked pintail, black-necked stilt, royal tern, Caribbean elaenia. **MAY–AUGUST**: laughing gull, Sandwich tern, least tern. **SEPTEMBER–APRIL**: Wilson's plover, greater and lesser yellowlegs, red knot, semi-palmated sandpiper, western sandpiper, least sandpiper, white-rumped sandpiper, pectoral sandpiper, barn swallow. **RARITIES**: Wilson's phalarope, black tern.

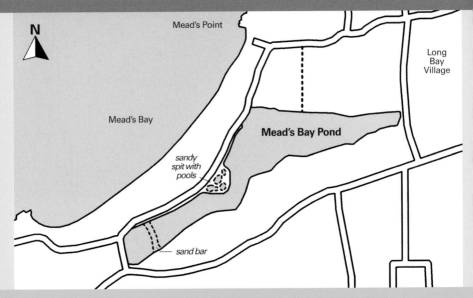

Mead's Bay Pond

ACCESS Leave the ANT building turning left to the roundabout and left following the main road to the west of the island for 7.1 miles. After passing the petrol station on the left, you continue west to the bottom of the hill and turn right. Travelling down the hill, Mead's Bay Pond can be seen on the right hand side. There are opportunities to view the pond along its full length. The pond is in public ownership.

VIEWING The pond is one of the deepest on the island, a result of sand dredging in the 1980s. Parts of the pond are now over three feet deep, sufficient to maintain a large water area year round. A sandy spit in the middle and a bund on the western side are also remnants from the sand dredging operations. The pond is now cut off from the bay by a belt of dunes.

Herons, egrets and ducks can be found around the distant rocky edges of the pond and pelicans will often drop in to feed or rest. The sandy shore by the road attracts shore birds and a shallow bay at the western end and a muddy spit by the road in the centre can be particularly good if there are pools of water present. When water levels are falling the pond attracts large numbers of shore birds such as yellowlegs and the smaller sandpipers. In the breeding season, Sandwich and least terns will feed in the pond and are occasionally joined by roseate and common terns. The central spit attracts breeding plovers and occasionally least terns, and should be viewed from the roadside to avoid disturbance. The western end of the pond has a range of salt meadow plants including torrey and a range of grasses.

The dune system attracts land birds and is a good spot for the American kestrel. The bay often holds shoals of small fish attracting brown boobies, terns and laughing gulls. Late afternoon is a good time to be on the beach with sanderlings running between the waves and yellow-crowned night-herons, green herons and belted kingfishers flying onto the rocks at dusk on the eastern end of the bay. Nighthawks are occasionally seen hawking over the headland in the early evening. The mature gardens of the Mallihouana Hotel attract most of the island's land birds and are a good area to look for scarce North American warblers.

WHAT TO SEE YEAR ROUND: snowy egret, green heron, white-cheeked pintail, American kestrel, black-necked stilt, gray kingbird, black-faced grassquit, Lesser Antillean bullfinch. **MAY–AUGUST**: snowy plover, Wilson's plover, laughing gull, Sandwich tern, least tern, Antillean nighthawk. **SEPTEMBER–APRIL**: pied-billed grebe, tri-coloured heron, yellow-crowned night-heron, greater and lesser yellowlegs, sanderling, belted kingfisher, barn swallow. **RARITIES**: little blue heron, ruff, scaly-naped pigeon, chimney swift, black-throated green warbler.

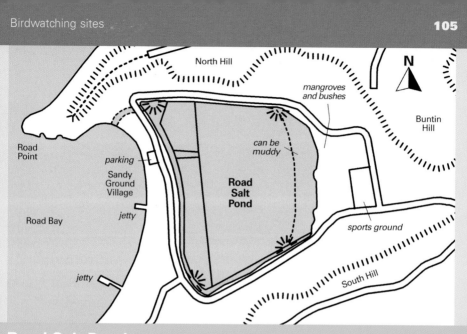

Road Salt Pond

ACCESS Leave the ANT building turning left to the roundabout following the main road to the west of the island from the roundabout near the airport. Continue west until the next roundabout and take the right turn and go down the hill. This road provides spectacular views of the pond and bay at Sandy Ground. At the bottom of the hill, take the right turn at the sports field. From this point, there is a circular walk or drive around the entire pond although the busy road is hard against the pond at its western side. The pond area is in public ownership.

VIEWING This is a spectacular pond and at peak times a walk around its perimeter will provide the most varied birdwatching on the island. The pond covers 98 acres (39.6 ha) and until the 1980s was a centre of salt production. The salt pond walls remain largely intact and there are other remains of the industry in Sandy Ground village. The pond retains water year round but in prolonged dry periods the areas outside the salt pond walls and the inland eastern edge dry out. A mixture of shallow water and muddy edges is perfect for migratory and resident water birds and large flocks of lesser yellowlegs and stilt sandpipers can gather to feed on brine flies and other prey items with over 900 birds present in October 2006. The landward edge and the area nearest Road Bay often afford close views of a wide range of species. Check grassy areas on the bay side for more unusual species such as Wilson's snipe and pectoral sandpipers. After feeding, gulls, terns, ducks and shore birds will rest on the long salt pond walls.

Herons and egrets will roost in the bushes and mangroves on the eastern edge. Cattle egrets and killdeers feed on the nearby sports field and a variety of land birds can be found around the quieter northern edge of the pond including the mangrove cuckoo. An inlet to the sea at the north-east corner is good for green herons and feeding least terns. The large numbers of birds usually present attracts peregrines, merlins and ospreys can often be seen here if present on the island. The bay is good for a range of seabirds. Frigatebirds are a regular sight over Road Point and it is worth checking this point at dusk in April–August for feeding Antillean nighthawks.

What to see YEAR ROUND: magnificent frigatebird, green heron, yellow-crowned night heron, white-cheeked pintail, common moorhen, killdeer, mangrove cuckoo, Caribbean elaenia, gray kingbird. **MAY–AUGUST:** laughing gull, Sandwich tern, least tern, Antillean nighthawk. **SEPTEMBER–APRIL:** great egret, blue-winged teal, osprey, merlin, peregrine falcon, sora rail, spotted sandpiper, ruddy turnstone, semi-palmated sandpiper, least sandpiper, white-rumped sandpiper, pectoral sandpiper, stilt sandpiper, short-billed dowitcher, Wilson's snipe, royal tern, yellow-billed cuckoo, blackpoll warbler. **RARITIES:** hooded merganser, little blue heron, glossy ibis, Hudsonian godwit, great black-backed gull, scarlet tanager.

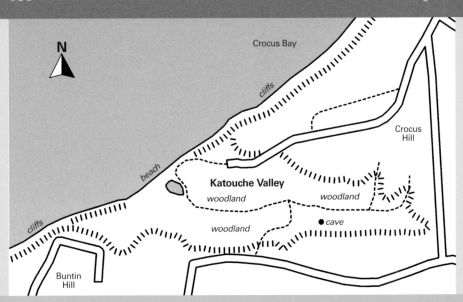

Katouche Valley and Pond

ACCESS Leave the ANT office and turn left. Go straight across at the roundabout heading out towards Crocus Bay. After 0.8 miles take a left turn with the Red Cross building opposite on your right. Follow this road for 0.5 miles to the end where it bears right and winds its way down a steep hill to Katouche Bay. Walk down to the beach and turn left. After a few yards follow the track through the bushes to Katouche Pond. A track on the western side of the pond leaves along the valley bottom and after several hundred yards splits with the right hand track climbing the side of the valley to a rock track. This is a privately owned site. If you find access difficult, please contact the ANT.

VIEWING The walk through the valley is a wonderful experience taking you under the canopy of the best remaining area of dry evergreen forest on the island. The humidity changes within the woodland and the rich variety of trees includes wild fig, loblolly, turpentine and several rare species resulting in a bewildering array of beautiful leaf shapes. The forest also includes a range of ferns and fungi. As you walk through the valley look out for Anguilla tree lizards, butterflies, hermit crabs and the occasional red-footed tortoise.

Bird songs and calls can fill the valley but it can be difficult to see birds in the dense canopy. The pearly-eyed thrasher is probably at its highest density on the island. Hummingbirds and bananaquits are usually present and the valley is a regular haunt and wintering area for North American land birds although these can be hard to find.

The access to the small pond is difficult through a thick stand of bushes and care is needed as there are a few poisonous manchineel bushes near the beach although the pond edges are now lined with mangrove and buttonwood trees. The pond is seasonal and will dry out but when full, large pond crabs are found around the pond and in holes among the bushes. Branches of dead trees around the pond provide perches for herons and a few ducks. Shore birds are also usually present and the belted kingfisher is an occasional visitor. The area around the pond is a good start point in the search for warblers.

Katouche Bay is part of a stretch of coastline attracting a range of seabirds especially brown pelicans and red-billed tropicbirds, which have been seen around the sandstone cliffs.

WHAT TO SEE YEAR ROUND: yellow-crowned night-heron, green-throated carib, Antillean crested hummingbird, Caribbean elaenia, pearly-eyed thrasher, yellow warbler. **SEPTEMBER–APRIL**: belted kingfisher, lesser yellowlegs, spotted sandpiper, northern parula, prairie warbler, blackpoll warbler, black-and-white warbler, northern waterthrush. **RARITIES**: magnolia warbler, yellow-rumped warbler, black-throated green warbler, Prothonotary warbler, Swainson's warbler.

Other sites

Blackgarden Bay Pond

This is a tiny brackish pond to the east of Limestone Bay. Access to the privately owned area is difficult but it can be overlooked from the western side. A sand bar separates it from the sea and the pond is surrounded by mangroves and manchineel trees. The slope on the eastern side is attractively covered in a range of mature trees.

The pond is a regular year-round haunt of the yellow-crowned night-heron, mangrove cuckoo and pearly-eyed thrasher. A few shore birds and belted kingfisher are regular visitors in the October–March period.

Brimegin

This is one of the best and most extensive areas of limestone scrub running from Limestone Bay east to Shoal Bay Village. The area is in private ownership and the only current development is an extensive area of rock quarrying. The scrub extends to the coast where sea cliffs rise to around 15 metres (50 feet) and this area is one of the last refuges for the native iguanas. Access is difficult with few tracks. The scrub holds good populations of most land birds and the cliffs and coast are one of the best areas to see brown boobies, brown pelicans and American oystercatchers.

Long Bay Pond

This small seasonal pond of 1.7 acres lies just behind the beach at Long Bay east of Mead's Bay on the north coast. The privately owned pond can be seen from the access road to a new development on its western side. The pond has some mangroves although these were badly damaged by Hurricane Luis in 1995. The pond attracts a few wetland birds including white-cheeked pintails and shore birds in the period September–April.

Temenos golf course and Merrywing Pond

This new, private development is in an area of former scrub east of Merrywing Pond. The pond itself has been incorporated into the development and its margins have been heavily modified. The golf course has an irrigation system with holding ponds and a stream running through a series of modified wetland habitats near the coast. The fairways provide extensive areas of irrigated grasslands and the whole complex attracts a range of water birds and land birds. Merrywing Pond can be overlooked from the eastern side of Cove Bay Pond and the ANT is in touch with the site managers over access agreements to the new wetland areas. The fairways are particularly good for cattle egrets, American golden plovers (September–November), killdeers and pectoral sandpipers (August–November). The wetlands attract a range of herons, egrets, ducks and shore birds and will be particularly good when seasonal ponds dry out.

West End Point and Anguillita Island

This area lies to the west of West End village and is in public ownership. Access is difficult through the scrub although there are paths kept open by fishermen and goats. There is a coastal path in parts leading to some attractive limestone and sandstone cliffs particularly on the northern shore. In the central part there is an area of dry forest and extensive areas of scrub holding a range of land birds. If you are able to reach the tip at Lower West End point the cliffs are good for brown pelicans and a few brown noddies have been seen on the cliffs here in August.

From the point you can look south-west to the low rocky shelf of Anguillita Island. The island has a sand bar at its centre and attracts breeding terns including royal terns. The island may also provide a night-time roost for shore birds as several parties have been seen flying out towards the island in the late afternoon.

West End by-the-sea Pond

This small pond is at the western end of West End Bay and is reached by continuing straight on when you first arrive at West End Pond. The pond is on a track to the right and lies behind the beach. The area is split by the track with dense areas of mangrove and buttonwood on the eastern side and more emergent vegetation on the west. The pond is seasonal but when full can attract small numbers of a range of wetland birds including yellow-crowned night-herons. The mangroves and bushes are known to attract warblers such as the northern waterthrush and can be a good area to see a yellow warbler.

The smaller islands and cays

Dog Island

This privately owned island of 512 acres (207 ha) lies 9 miles off the mainland. It shares many of the natural features of Anguilla with limestone and sandstone cliffs, long sweeps of beach, limestone scrub and two coastal lagoons. In common with all the smaller islands and cays it is uninhabited although there are remains of agriculture with low walls and a small herd of goats. The centre of the island is covered in impenetrable low thorny bushes interspersed with prickly pear cacti. Tiny rock outcrops occur off the main island at East, Mid and West Cay. The best way to experience the island is by boat and all the seabirds can be seen on and around the coastal strip.

Recent bird surveys have shown the island to be a natural treasure and one of the best seabird breeding islands in the Caribbean. The scrub is particularly good for sooty terns and the colony may reach 100,000 birds. Ten species of seabird will breed here including the masked booby, it holds Anguilla's largest population of brown booby (over 1,250 pairs) and its only breeding magnificent frigatebirds. The main breeding season for the terns runs from April to July. There is concern that rats, recently found on the island, will pose a significant threat to the seabirds. The ponds and coast attract wetland and shore birds. Nesting green and hawksbill turtles use the remote sandy beaches.

Little Scrub Island

A small publicly owned island to the west of the much larger Scrub Island, it is 11 acres (4.5 ha) in area and is a rock outcrop with little vegetation. The island has breeding populations of terns, especially brown noddy. The island is famous for its lizard populations including the endemic Little Scrub ground lizard.

Prickly Pear East

This privately owned island lies 5 miles to the north of Anguilla and covers 76 acres (31 ha). It is a low, rocky island with sandy shores on the north and east coasts and heavily fissured limestone on the remaining coast. A small pond lies behind the northern beach and the centre of the island has areas of scrub. The island has inshore coral reefs that have been a popular destination for visitors from the mainland and St Martin. A restaurant has been built that opens at peak visitor times making this the most accessible and most visited small island.

The island holds populations of breeding seabirds including brown boobies, bridled terns, brown noddies and a large colony of laughing gulls. These species are all vulnerable to disturbance and if you visit the island, care should be taken not to venture into the breeding colonies. The pond attracts small numbers of water birds and the scrub holds several species of land bird including Caribbean elaenias and yellow warblers.

Prickly Pear West

This small island of 79 acres (32 ha) is separated by a narrow channel from Prickly Pear East and is a rockier more rugged island. Apart from one small beach, the island is a low limestone outcrop with low cliffs. The centre is dominated by dense impenetrable scrub except for a narrow strip of bare rock that is home to a colony of brown boobies. The bushes in the centre provide an occasional nesting area for a few pairs of red-footed boobies at their only breeding site on Anguilla.

Scrub Island

At 847 acres (343 ha), this is the largest and most varied of the islands. It lies just off Windward Point and is another privately owned, and uninhabited, island although there are remains of tourism development and a disused airstrip. The island has a mixture of beautiful sandy beaches, ponds and coastal lagoons and low rocky shores and cliffs. The island remains little visited.

The coastal lagoons of Deadmans Bay in the east are home to large colonies of terns in the period April–August and include a regionally important colony of roseate terns. These lagoons and the pond at Scrub Bay hold a range of breeding and migratory water birds including willets. The secluded coasts are also home to a noisy colony of laughing gulls and a population of American oystercatchers. The hinterland of the island includes large areas of scrub and has populations of most land birds and a range of reptiles including the Anguillian racer snake.

Sombrero

Set on its own rocky bank some 40 miles to the north-west of the mainland, Sombrero is a remote rocky outpost of 95 acres (38 ha), famous for its lighthouse and its seabird colonies. The lighthouse is now automatic leaving the publicly owned island uninhabited save for occasionally visiting biologists, fishermen and sailors. The island was formerly mined for phosphate and this has left much industrial archaeology and the rocky surface heavily fissured and pitted with craters. Sombrero hit the headlines in 1999 when a proposal came forward to build a satellite launch station there, but the application was subsequently withdrawn. There are now plans to declare the island a protected area.

Sombrero has a large colony of brown and masked boobies and from April large numbers of terns arrive including an important population of bridled terns together with sooty terns and brown noddies. In 1988 two black noddies, very rare in the West Indies, were found among the breeding terns. The island is home to the endemic black lizard and recent research has found a number of insects also found nowhere else in the world.

The island has been little studied during the bird migration seasons, when Sombrero's remote location has provided a landfall and refuge for many migratory water birds and land birds including several species of North American wood-warbler.

Checklist of the birds of Anguilla

At least 132 species of bird have been recorded on Anguilla since 1990. Around 88 species are either resident or are recorded each year and the rest are rare visitors, many of which have only been seen on a single occasion. Anguilla is well placed at the north-east corner of the Lesser Antilles to attract rare birds on migration or birds displaced by weather systems. The list of birds reported on neighbouring islands suggests the list of birds on Anguilla will continue to grow.

This checklist has been compiled from bird surveys and local sightings reported to the Anguilla National Trust (ANT) since 1990. Within this period of 17 years the number of active birdwatchers on the island has varied but in general the island's birds and other wildlife have been little studied. This situation has changed recently with a partnership between the ANT and the RSPB in the UK ensuring surveys of breeding seabirds and a series of wetland bird counts. Our knowledge of the status of birds on the islands is largely based on these surveys and increased levels of birdwatching since late 1999.

A small number of species are known to have occurred on the islands from verbal reports from local birdwatchers. The majority of these pre-date 1990 and unfortunately we do not have dates or further details for some of the records. These include:

Audubon's shearwater	*Puffinis lherminieri*
Black-crowned night-heron	*Nycticorax nycticorax*
Cinammon teal	*Anas cyanoptera*
Clapper rail	*Rallus longirostris*
Gull-billed tern	*Sterna nilotica*
Black noddy	*Anous minutus*
American robin	*Turdus migratorius*
Cape may warbler	*Dendroica tigrina*
Worm-eating warbler	*Helmithiros vermivorus*

Where possible all the reports in this checklist have been checked for accuracy and since 2001 most additions to the checklist have been the result of written notes, often supported by photographs, submitted to the ANT and checked. Where identification was uncertain some records have been left off the checklist to ensure it is as accurate as possible.

The checklist has been designed to provide an indication of how likely you are to see these species in each month of the year. The key to the bars in the monthly table are as follows:

- ■ = You should see this species if you visit the habitats they are found in.
- ■ = You should see this species if you visit favoured sites or search in suitable habitats. Some resident species have been placed in this category as they are elusive and can be difficult to find.
- ▨ = You could see these species although they are irregular visitors. Some species may occur each year, others have been seen on only one occasion. The column on status will provide a clearer indication of how often the species occurs.

No bar You are unlikely to see this species as there are no records in that month.

The status of birds on Anguilla is becoming clearer and is based on the following definitions which apply to the majority of birds seen on the islands as some species can include birds from more than one category:

- **Breeding resident** All, or the main part, of the population is present year round and the species breeds on the island.
- **Non-breeding resident** The species is present in varying numbers year round but does not breed. The population may include migrants too.
- **Breeding visitor** The species is present in the breeding season only.
- **Regular visitor** The species visits each year on migration although numbers may vary from year to year.

–   **Scarce visitor** The species visits in small numbers most years but not every year.

–   **Rare visitor** The species is not present every year and only visits occasionally or rarely. This category includes some species that are well away from usual range or migration route. Some of these birds will spend several months on the island and all months of presence are included in the checklist.

Note: The current status of resident and breeding birds is based on increased data gathered in recent years and is considered to be accurate. The status of migrants is based on more limited data and the status of several species is likely to change with several species likely to occur more frequently than current records suggest. The weather during periods of migration will continue to influence status within a particular year as the presence or absence of storms during migration has a direct effect on the numbers of migrating birds making landfall on Anguilla. It is hoped this guide will encourage more local people and visitors to send details of their sightings to the ANT helping to improve our knowledge of the birds and other wildlife on Anguilla.

Species order in the checklist

To help cross-referencing we have followed the common and scientific names, and systematic order in *Birds of the West Indies* by Rafaelle *et al.*

How you can help

The information recorded within this checklist is held by the Anguilla National Trust. The ANT would appreciate information on your bird sightings and these can now be added on-line through **www.worldbirds.org**. In particular if you can fill in any monthly gaps of presence or see any scarce or rare migrants please pass details to the Trust. In recent years detailed descriptions with photographs where possible are being collected for all rare migrants and we plan to follow this route for all new additions to the checklist. Alternatively if you see a species that is rare or new to Anguilla please contact the ANT office as soon as possible to enable someone to support your identification. We can include your sighting in future guides and publications!

Species	Status	J	F	M	A	M	J	J	A	S	O	N	D
Pied-billed grebe *Podilymbus podiceps*	Regular visitor	■	■	■	■	■	■	■		■	■	■	■
White-tailed tropicbird *Phaethon lepturus*	Breeding visitor			■	■	■	■	■					
Red-billed tropicbird *Phaethon aethereus*	Breeding resident	■	■	■	■	■	■	■	■	■	■	■	■
Masked booby *Sula dactylatra*	Breeding resident	■	■	■	■	■	■	■	■	■	■	■	■
Brown booby *Sula leucogaster*	Breeding resident	■	■	■	■	■	■	■	■	■	■	■	■
Red-footed booby *Sula sula*	Breeding visitor						■	■	■	■			
Brown pelican *Pelecanus occidentalis*	Breeding resident	■	■	■	■	■	■	■	■	■	■	■	■
Magnificent frigatebird *Fregata magnificens*	Breeding resident	■	■	■	■	■	■	■	■	■	■	■	■
Great blue heron *Ardea herodias*	Regular visitor	■	■	■	■	▪	▪				■	■	■

Species	Status	J	F	M	A	M	J	J	A	S	O	N	D
Great egret *Ardea alba*	Non-breeding resident	■	■	■	■	■	■	■	■	■	■	■	■
Snowy egret *Egretta thula*	Non-breeding resident	■	■	■	■	■	■	■	■	■	■	■	■
Little blue heron *Egretta caerulea*	Rare visitor	■	■	■	■	■	■						
Tri-coloured heron *Egretta tricolor*	Scarce visitor	■	■								■	■	■
Cattle egret *Bubulcus ibis*	Non-breeding resident	■	■	■	■	■	■	■	■	■	■	■	■
Green heron *Butorides virescens*	Breeding resident	■	■	■	■	■	■	■	■	■	■	■	■
Yellow-crowned night-heron *Nyctanassa violacea*	Breeding resident	■	■	■	■	■	■	■	■	■	■	■	■
Glossy ibis *Plegadis falcinellus*	Rare visitor											■	
Greater flamingo *Phoenicopterus ruber*	Rare vistor	■										■	■
West Indian whistling duck *Dendrocygna arborea*	Rare visitor												■
Snow goose *Chen caerulescens*	Rare visitor	■	■									■	■
Green-winged teal *Anas crecca*	Scarce visitor	■	■	■							■	■	■
White-cheeked pintail *Anas bahamensis*	Breeding resident	■	■	■	■	■	■	■	■	■	■	■	■
Northern pintail *Anas acuta*	Rare visitor	■	■										
Blue-winged teal *Anas discors*	Regular visitor	■	■	■	■						■	■	■
Northern shoveler *Anas clypeata*	Rare visitor	■	■										
American wigeon *Anas americana*	Scarce visitor	■										■	■
Ring-necked duck *Aythya collaris*	Scarce visitor	■	■	■	■							■	
Lesser scaup *Aythya affinis*	Scarce visitor	■	■	■	■							■	■
Hooded merganser *Lophodytes cucullatus*	Rare visitor	■	■										
Ruddy duck *Oxyura jamaicensis*	Regular visitor	■	■	■	■	■	■	■	■	■	■	■	■

Species	Status	J	F	M	A	M	J	J	A	S	O	N	D
Osprey *Pandion haliaetus*	Regular visitor	■	■	■	■	▪	▪			▪	■	■	■
Broad-winged hawk *Buteo platypterus*	Rare visitor			▪									
American kestrel *Falco sparverius*	Breeding resident	■	■	■	■	■	■	■	■	■	■	■	■
Merlin *Falco columbarius*	Regular visitor	■	■	■	■						■	■	■
Peregrine falcon *Falco peregrinus*	Regular visitor	■	■	■							■	■	■
Sora rail *Porzana carolina*	Scarce visitor	■	■								■	■	■
Common moorhen *Gallinula chloropus*	Breeding resident	■	■	■	■	■	■	■	■	■	■	■	■
American coot *Fulica americana*	Breeding resident	■	■	■	■	■	■	■	■	■	■	■	■
Caribbean coot *Fulica caribaea*	Breeding visitor	■	■	■	■	■	■						■
Black-bellied plover *Pluvialis squatarola*	Regular visitor	■	■	■	■		▪	▪	▪	▪	■	■	■
American golden plover *Pluvialis dominica*	Scarce visitor									■	■		
Snowy plover *Charadrius alexandrinus*	Breeding resident	■	■	■	■	■	■	■	■	■	■	■	■
Wilson's plover *Charadrius wilsonia*	Breeding resident	■	■	■	■	■	■	■	■	■	■	■	■
Semi-palmated plover *Charadrius semipalmatus*	Regular visitor	■	■	■	■	■	■	■	■	■	■	■	■
Piping plover *Charadrius melodus*	Rare visitor	▪								▪	▪	▪	
Killdeer *Charadrius vociferus*	Breeding resident	■	■	■	■	■	■	■	■	■	■	■	■
American oystercatcher *Haematopus palliatus*	Breeding resident	■	■	■	■	■	■	■	■	■	■	■	■
Black-necked stilt *Himantopus mexicanus*	Breeding resident	■	■	■	■	■	■	■	■	■	■	■	■
American avocet *Recurvirostra americana*	Rare visitor						▪						
Greater yellowlegs *Tringa melanoleuca*	Regular visitor	■	■	■	■	▪	▪	▪	▪	▪	■	■	■
Lesser yellowlegs *Tringa flavipes*	Regular visitor	■	■	■	■	■	■	■	■	■	■	■	■

Species	Status	J	F	M	A	M	J	J	A	S	O	N	D
Solitary sandpiper *Tringa solitaria*	Scarce visitor								▪	▪	▪	▪	
Willet *Catoptrophorus semipalmatus*	Breeding visitor	▪		▪	▪	▪	▪	▪	▪	▪			
Spotted sandpiper *Actitis macularia*	Regular visitor	■	■	■	■	▪		▪		■	■	■	■
Whimbrel *Numenius phaeopus*	Regular visitor	▪	▪	▪	▪	▪			▪	▪	▪	▪	▪
Hudsonian godwit *Limosa haemastica*	Rare visitor									▪			
Ruddy turnstone *Arenaria interpres*	Regular visitor	■	■	■	■	■	▪		▪	■	■	■	■
Red knot *Calidris canutus*	Scarce visitor	▪						▪	▪				▪
Sanderling *Calidris alba*	Regular visitor	■	■	■	■	▪			▪	■	■	■	■
Semi-palmated sandpiper *Calidris pusilla*	Regular visitor	■	■	■	■	■		▪	■	■	■	■	■
Western sandpiper *Calidris mauri*	Scarce visitor			▪				▪			▪	▪	
Least sandpiper *Calidris minutilla*	Regular visitor	■	■	■	■	■		▪	■	■	■	■	■
White-rumped sandpiper *Calidris fuscicollis*	Regular visitor	▪	▪						■	■	■	■	
Pectoral sandpiper *Calidris melanotos*	Regular visitor								■	■	■	■	
Curlew sandpiper *Calidris ferruginea*	Rare visitor						▪						
Stilt sandpiper *Calidris himantopus*	Regular visitor	■	■	■	■	■		▪	▪	■	■	■	■
Ruff *Philomachus pugnax*	Rare visitor			▪									
Long-billed dowitcher *Limnodromus scolopaceus*	Rare visitor										▪		
Short-billed dowitcher *Limnodromus griseus*	Regular visitor	■	■	■	■	■	■	■	■	■	■	■	■
Wilson's snipe *Gallinago delicata*	Regular visitor	▪	▪	▪						▪	▪	▪	▪
Wilson's phalarope *Phalaropus tricolor*	Scarce visitor									■			
Laughing gull *Larus atricilla*	Breeding visitor	▪			■	■	■	■	■	▪	▪	▪	

Species	Status	J	F	M	A	M	J	J	A	S	O	N	D
Ring-billed gull *Larus delawarensis*	Rare visitor		■		■								
Lesser black-backed gull *Larus fuscus*	Rare visitor										■		
Great black-backed gull *Larus marinus*	Rare visitor					■							■
Caspian tern *Sterna caspia*	Rare visitor					■							
Royal tern *Sterna maxima*	Breeding resident	■	■	■	■	■	■	■	■	■	■	■	■
Sandwich tern *Sterna sandvicensis*	Breeding visitor				■	■	■	■	■	■	■		
Roseate tern *Sterna dougallii*	Breeding visitor				■	■	■	■	■				
Common tern *Sterna hirundo*	Regular visitor				■	■	■	■	■	■	■	■	
Least tern *Sterna antillarum*	Breeding visitor			■	■	■	■	■	■	■			
Bridled tern *Sterna anaethetus*	Breeding visitor				■	■	■	■	■				
Sooty tern *Sterna fuscata*	Breeding visitor				■	■	■	■	■				
Black tern *Chlidonias niger*	Rare visitor										■		
Brown noddy *Anous stolidus*	Breeding visitor				■	■	■	■					
Rock dove *Columba livia*	Breeding resident	■	■	■	■	■	■	■	■	■	■	■	■
Scaly-naped pigeon *Columba squamosa*	Rare visitor			■									
White-crowned pigeon *Columba leucocephala*	Rare visitor							■					
Eurasian collared dove *Stretopelia decaocto*	Rare visitor					■			■				
White-winged dove *Zenaida asiatica*	Non-breeding resident	■	■	■	■	■	■	■	■	■	■	■	■
Zenaida dove *Zenaida aurita*	Breeding resident	■	■	■	■	■	■	■	■	■	■	■	■
Common ground-dove *Columbina passerina*	Breeding resident	■	■	■	■	■	■	■	■	■	■	■	■
Yellow-billed cuckoo *Coccyzus americanus*	Regular visitor										■	■	

Species	Status	J	F	M	A	M	J	J	A	S	O	N	D
Mangrove cuckoo *Coccyzus minor*	Breeding resident	■	■	■	■	■	■	■	■	■	■	■	■
Common nighthawk *Chordeiles minor*	Scarce visitor									■			
Antillean nighthawk *Chordeiles gundlachii*	Breeding visitor			■	■	■	■	■	■	■			
Chimney swift *Chaetura pelagica*	Rare visitor									■			
Purple-throated carib *Eulampsis jugularis*	Rare visitor									■			
Green-throated carib *Eulampsis holosericeus*	Breeding resident	■	■	■	■	■	■	■	■	■	■	■	■
Antillean crested hummingbird *Orthorhyncus cristatus*	Breeding resident	■	■	■	■	■	■	■	■	■	■	■	■
Belted kingfisher *Ceryle alcyon*	Regular visitor	■	■	■	■						■	■	■
Caribbean elaenia *Elaenia martinica*	Breeding resident	■	■	■	■	■	■	■	■	■	■	■	■
Gray kingbird *Tyrannus dominicensis*	Breeding resident	■	■	■	■	■	■	■	■	■	■	■	■
Caribbean martin *Progne dominicensis*	Scarce visitor				■	■		■					
Tree swallow *Tachycineta bicolor*	Rare visitor											■	■
Bank swallow *Riparia riparia*	Scarce visitor	■									■		■
Barn swallow *Hirundo rustica*	Regular visitor	■	■	■	■	■			■	■	■	■	■
Northern mockingbird *Mimus polyglottos*	Rare visitor										■		
Pearly-eyed thrasher *Margarops fuscatus*	Breeding resident	■	■	■	■	■	■	■	■	■	■	■	■
Northern parula *Parula americana*	Scarce visitor	■	■		■						■	■	■
Yellow warbler *Dendroica petechia*	Breeding resident	■	■	■	■	■	■	■	■	■	■	■	■
Magnolia warbler *Dendroica magnolia*	Rare visitor			■									
Black-throated blue warbler *Dendroica caerulescens*	Rare visitor				■								
Yellow-throated warbler *Dendroica dominica*	Rare visitor	■											

Species	Status	J	F	M	A	M	J	J	A	S	O	N	D
Yellow-rumped warbler *Dendroica coronata*	Rare visitor		■		■							■	
Black-throated green warbler *Dendroica virens*	Rare visitor	■			■						■		
Prairie warbler *Dendroica discolor*	Regular visitor	■	■								■	■	■
Blackpoll warbler *Dendroica striata*	Regular visitor									■	■		
Black-and-white warbler *Mniotilta varia*	Regular visitor	■	■								■	■	■
Prothonatary warbler *Protonotaria citrea*	Rare visitor			■									
Ovenbird *Seiurus aurocapillus*	Rare visitor										■		
Northern waterthrush *Seiurus noveboracensis*	Regular visitor	■	■	■							■	■	■
Swainson's warbler *Limnothylypis swainsonii*	Rare visitor	■											
Common yellowthroat *Geothylpis trichas*	Rare visitor											■	■
Hooded warbler *Wilsonia citrina*	Rare visitor	■	■								■		
Bananaquit *Coereba flaveola*	Breeding resident	■	■	■	■	■	■	■	■	■	■	■	■
Antillean euphonia *Euphonia musica*	Rare visitor	■											
Scarlet tanager *Piranga olivacea*	Rare visitor										■		
Black-faced grassquit *Tiaris bicolor*	Breeding resident	■	■	■	■	■	■	■	■	■	■	■	■
Lesser Antillean bullfinch *Loxigilla noctis*	Breeding resident	■	■	■	■	■	■	■	■	■	■	■	■
Shiny cowbird *Molothrus bonariensis*	Rare visitor			■									
House sparrow *Passer domesticus*	Breeding resident	■	■	■	■	■	■	■	■	■	■	■	■

The scientific names of plants and animals mentioned in the text

Plants

Bromeliad/Wild pine	*Tillandsia utriculata*
Spike rush	*Eleocharis mutate*
Seashore papsalum	*Papsalam vaginatum*
Seashore rush grass	*Sporobolus virginicus*
Pondweed (sea purslane)	*Sesuvium portulacastrum*
Frangipani	*Plumeria alba*
Licewood	*Urechite lutea*
Black mangrove	*Avicennia germinans*
Cedar	*Tabebuia heterphylla*
Sea lavender (wild bay)	*Argusia gnapholodes*
Turpentine tree	*Bursera simaruba*
Turk's head cactus (Pope's head)	*Melocactus intortus*
Prickly pear cactus	*Opuntia dillenii*
Doodle doo cactus	*Pilosocereus royeni*
Torrey (glasswort)	*Salicornia bigelovii*
Buttonwood	*Conocarpus erectus*
White mangrove	*Laguncularia racemosa*
Red mangrove	*Rhizophora sp.*
Lavender (sea marigold)	*Borrichia arborescens*
Sea bean (beach morning glory)	*Ipomoea pes-caprae*
Seaside spurge	*Chamaesyce mesembrianthemifolia*
Manchineel	*Hippomane mancinella*
Tamarind	*Tamarindus indica*
Cow bean	*Canavalia rosea*
Acacia	*Acacia macracantha*
Mahogany	*Swietenia mahogoni*
Wild fig	*Ficus citrifolia*
Loblolly	*Pisonia subcordata*
Sea grape	*Coccoloba uvifera*
Pomme-surette	*Ziziphus mauritiana*
Thorn	*Ziziphus rignonii*
Anguilla bush	*Rondaletia anguillensis*
Rosemary	*Strumpfia maritime*
Sage cop	*Lantana involucrate*

Reptiles and amphibians

Iguana	*Iguana delicatissima*
Green iguana	*Iguana iguana*
Ground lizard	*Ameiva plei*
Little Scrub ground lizard	*Ameiva corax*
Sombrero ground (black) lizard	*Ameiva corvine*
Tree lizard	*Anolis gingivinus*
Anguilla bank racer	*Alsophilus rijgersmaei*
Green turtle	*Chelonia mydas*
Hawksbill turtle	*Eretmochelys imbricate*
Leatherback turtle	*Dermochelys coriacea*
Red-footed tortoise	*Geochelone carbonaria*

References

The following books and publications were referred to in the writing of this guide:

ANT (1998) *A field guide to Anguilla's wetlands*. Anguilla National Trust, Anguilla.

Bond J (1960) *The birds of the West Indies*. Collins, UK.

Bryer M, Fisher I, Holliday S H and Hughes J (2000) *Birds of Anguilla and outer islands*. The RSPB, Sandy, UK.

Collier N and Bown A C (2004) *Environmental protection in the Caribbean, report no 22*. Epic Islands report to the Anguilla National Trust, Anguilla.

Evans P G H (1990) *Birds of the eastern Caribbean*. Macmillan Education Ltd, London.

Hodge K V D, Censky E J and Powell R (2003) *The reptiles and amphibians of Anguilla, British West Indies*. Anguilla National Trust, Anguilla.

Holliday S H and Hodge K V D (2006) Anguilla (pp 9–18) in S M Sanders, *Important Bird Areas in the United Kingdom Overseas Territories*. The RSPB, Sandy, UK.

ICF Consulting (1999) *Supplemental biological surveys on Sombrero island and other Anguillian islands*. EIA report.

Norton R L (1989) First West Indian record of black noddy and nesting masked booby at Sombrero Island, Lesser Antilles. *Colonial Waterbirds* 12(1): 120–122.

Pritchard D (1990) *The Ramsar Convention in the Caribbean: with special emphasis on Anguilla*. The RSPB, Sandy, UK.

Rafaelle H, *et al.* (1998) *Birds of the West Indies*. Helm, London.

Sibley D (2000) *The North American bird guide*. Pica Press, East Sussex

Thomas R (1996, revised 1997) *Birds of Anguilla*. Resortscapes: Woodstock, VT, USA.

Walker M M, Hodge O, Homer F and Johnson W (2005) *A guide to the common plants of Anguilla*. Anguilla National Trust, Anguilla.

Index

Common names are shown in **bold**, local names in roman and scientific names in *italic.*